# PRACTICAL PROBLEMS in MATHEMATICS

## for CARPENTERS

### Fourth Edition

## HARRY C. HUTH

**DELMAR PUBLISHERS inc.**

*Delmar Staff*

Administrative Editor:  Mark W. Huth
Production Editor:  Ruth Saur

For information, address Delmar Publishers Inc.
2 Computer Drive West, Box 15-015
Albany, New York  12212

Printed in the United States of America
Published simultaneously in Canada
by Nelson Canada,
A division of International Thomson Limited

10  9  8  7  6  5  4  3  2  1


**Library of Congress Cataloging in Publication Data**

Huth, Harry C.
Practical problems in mathematics for carpenters.

1. Carpentry—Mathematics.   I. Title.
TH5612.H87      1985      513′.132      84-12174
ISBN 0-8273-2427-8

# *CONTENTS*

## SECTION 6 RATIO AND PROPORTION / 125

## SECTION 7 POWERS AND ROOTS / 133

## SECTION 8 ESTIMATING / 144

# PREFACE

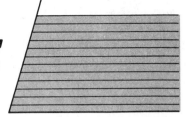

*Practical Problems in Mathematics for Carpenters* has been written to give students, who are interested in carpentry, practice in computing problems common to the carpentry trade. The student's comprehension of terminology used in the field of carpentry is strengthened through the use of this workbook.

The workbook is an excellent supplement to any vocational mathematics text. A chart cross-referencing the workbook to several popular mathematics and carpentry textbooks has been included in the Instructor's Guide in order to help instructors locate explanations of principles involved in solving the problems found in the workbook.

*Practical Problems in Mathematics for Carpenters* is one in a series of workbooks designed to offer students practical problem-solving experience in various occupations. The workbooks offer a step-by-step approach to the mastery of basic skills of mathematics. Each workbook includes relevant and easily understood problems in a specific vocational field. The workbooks are suitable for any student from the junior high school level through high school, and up to the two-year college level. Two Achievement Reviews are included at the end of each workbook to provide an effective means of measuring progress made by the student. A Diagnostic Reading Survey is also included. The survey will help determine the student's level of basic skill. A review of denominate numbers, a glossary, and answers to odd-numbered problems are also included at the end of the workbook.

The fourth edition of *Practical Problems in Mathematics for Carpenters* has brought material prices, interest rates, and labor costs used in the problems in line with today's economy.

The Instructor's Guide provides solutions to most of the problems and answers to all of the problems found in the workbook. Other instructional aids that will be helpful to the teacher are also included in the Instructor's Guide.

Harry C. Huth has been a teacher of mathematics for over 20 years. He is a graduate of Ithaca College, having studied at Rutgers University, Syracuse University and Hartwick College. He received his Master's Degree in Mathematics Education from Colgate University. In addition, Mr. Huth is an experienced carpenter of many years standing. He is also active in many educational associations.

# Whole Numbers

## Unit 1  ADDITION OF WHOLE NUMBERS

### REVIEW PROBLEMS

Add the following quantities:

1.  13 feet
    16 feet
     8 feet
    24 feet
    51 feet

4.  270 pounds
    140 pounds
    368 pounds
    609 pounds
    306 pounds

7.    261 square inches
    1,094 square inches
      861 square inches
      644 square inches
    1,326 square inches

2.  1 312 metre
      644 metre
       31 metre
    3 609 metre
      234 metre

5.  $1,608
      914
       83
    1,144

8.      23 grams
    1 116 grams
    2 492 grams
    3 844 grams

3.  3,002
      994
      681
    2,411

6.  $12,492
     2,063
     3,876
     9,432

9.  15,642 board feet
     8,431 board feet
    13,092 board feet
    15,842 board feet
     4,467 board feet

10.  In finishing off the ceiling trim for a paneled family room measuring 12′ x 16′, how many linear feet of cove molding are needed? _____

11.  As shown on the partial floor plan, what is the total length of the house ? _____

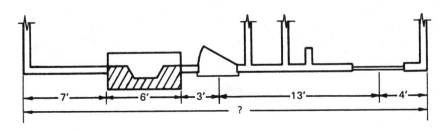

12. A basement has a recreation room, store room, and laundry. The floor spaces are 286 sq. ft., 91 sq. ft., and 164 sq. ft. respectively. What is the total number of square feet (sq. ft.) in this floor space?

_____

13. Four houses require 800 sq. ft., 1,050 sq. ft., 1,200 sq. ft., and 1,460 sq. ft. of asphalt shingles respectively. How many square feet (sq. ft.) of shingles are needed for the four houses?

_____

14. A crew of four worked 24 hours, 18 hours, 20 hours, and 22 hours. What is the total number of hours worked?

_____

15. A contractor paid bills of $2,480, $765, $1,446, $1,011, and $1,610 for materials. Find the total cost of the materials.

_____

16. In making five excavations, the following cubic metres of earth were removed: 5 040, 6 070, 7 940, 11 424, and 35 216. Find the total number of cubic metres of earth removed.

_____

17. Three deliveries of 1″ x 6″ roof boards are as follows: 2,450 board feet, 2,760 board feet, and 2,875 board feet. What is the total number of board feet delivered?

_____

18. A carpenter lays 1,300 wood shingles the first day, 1,400 the second, and 1,500 the third. How many shingles are used in three days?

_____

19. What is the total number of square feet of floor underlayment needed to complete a job if three rooms require 120 sq. ft., 90 sq. ft., and 300 sq. ft. respectively?

_____

20. A contractor buys 14,500 board feet of 1-inch white pine, 1,250 board feet of 2-inch spruce, and 1,450 board feet of 3-inch hemlock. Find the total number of board feet.

_____

21. The cost of material for a remodeling job is as follows: lumber, $476; masonry, $148; hardware, $62; and paint, $85. What is the total cost of materials?

_____

22. In estimating the finish flooring for a house, a contractor lists the room areas as follows: living room, 168 square feet; dining room, 152 square feet; bedroom, 142 square feet; hall, 45 square feet; and kitchen, 125 square feet. What is the total area to be floored?

_____

23. For a building, the following items of framing lumber are ordered: 472 board feet of 2″ x 4″ studs; 1,627 board feet of 2″ x 10″ joists; 827 board feet of 2″ x 6″ stock; 572 board feet of 2″ x 8″ stock. How many board feet of framing lumber are ordered?

_____

24. A residence measures 28′ x 54′. Plans show a 2″ x 6″ single sill plate. How many linear feet of sill plate are needed?

_____

**25.** The amounts of materials needed to construct the illustrated outbuilding are shown in the chart.  Determine the total number of board feet required for the job.

|  | Material | Quantity |
|---|---|---|
| A | Sill | 64 board feet |
| B | Plate Stock | 148 board feet |
| C | Studding | 380 board feet |
| D | Collar beams | 48 board feet |
| E | Rafters | 268 board feet |
| F | Headers | 40 board feet |
| G | Sheathing | 960 board feet |
|  | Total |  |

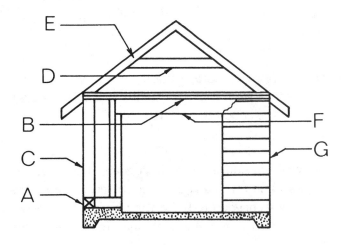

**26.** A contractor is building four buildings.  This is the order of supplies for the concrete block foundations.  Find the amount of each material needed for the total order.

| Foundation Sizes | a. Concrete Blocks | b. Masonry Cement | c. Sand |
|---|---|---|---|
| A  50′ x 32′ x  8′ | 1,476 blocks | 49 bags | 7 tons |
| B  24′ x 30′ x  8′ | 971 blocks | 32 bags | 5 tons |
| C  52′ x 36′ x 10′ | 1,979 blocks | 66 bags | 9 tons |
| D  20′ x 12′ x  9′ | 647 blocks | 23 bags | 3 tons |

Totals        a. _____        b. _____        c. _____

**Note:** The distance around a building is known as its *perimeter.* This measurement is often required by the carpenter, the mason, and other tradesmen in order to estimate quantities of materials needed. To find the perimeter of a building:

    a. For rectangular-shaped buildings, add the length and width together and double the result; or

    b. Add all the widths and lengths together.

27. Find the perimeters of the following rectangular-shaped buildings:

    a. 30'-0" x 45'-0"   _____      g. 29'-0" x 84'-0"   _____

    b. 52'-0" x 172'-0"   _____      h. 46'-0" x 49'-0"   _____

    c. 36'-0" x 102'-0"   _____      i. 51'-0" x 215'-0"   _____

    d. 49'-0" x 116'-0"   _____      j. 170'-0" x 60'-0"   _____

    e. 62'-0" x 214'-0"   _____      k. 35'-0" x 115'-0"   _____

    f. 48'-0" x 84'-0"   _____      l. 101'-0" x 45'-0"   _____

**Note:** To construct concrete sidewalks, garage floors, and driveways, it is often necessary to use a piece of 2" x 4" lumber (or some other suitable size), placed on edge, as a form to hold the concrete until it is set.

28. A garage floor and driveway are poured in a single slab. The driveway requires forms at both sides and on one end of the driveway. Find the number of linear feet of 2 x 4s needed to build the forms for the driveway and garage as shown. (Do not make allowance for joints at the corners of the forms.)

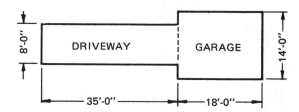

29. The outside forms for the footings of this house are one board high. Calculate the number of linear feet of 1″ x 8″ stock needed for the job. (Do not make any allowance for joints.)

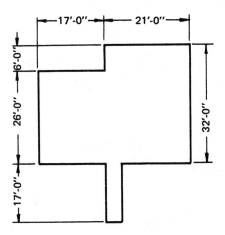

# Unit 2   SUBTRACTION OF WHOLE NUMBERS

## REVIEW PROBLEMS

Subtract the following:

1.  120 centimetres
    -  43 centimetres

4.  1,636
    -  703

7.  12,643 gallons
    -  7,123 gallons

10.  $14,254.00
     -  9,676.00

2.  1,473 pints
    -  611 pints

5.  3,469 square feet
    -  983 square feet

8.  16 298 kilograms
    -  9 696 kilograms

11.  55,864 board feet
     -  34,964 board feet

3.  909 ounces
    -  640 ounces

6.  2,091 yards
    -  993 yards

9.  23,122
    -  10,069

12.  $25,150.00
     -  11,909.00

**Note:** No allowance is made for the saw kerf.

13.  A length of stock 380 mm long is cut from a board 720 mm long. What is the length of the remaining piece?

14.  The supply of plywood in stock is 9,984 square feet. After 6,976 square feet are used, how much is left?

15.  From a supply of 7,326 linear feet of baseboard, a total of 4,560 feet is used. How many feet of baseboard remain in the supply?

16.  The area of a shop floor is 1,650 square feet. How much area remains to be painted after 735 square feet are covered?

17.  The excavator for the foundation of a house must remove 650 cubic metres of earth. How much remains to be excavated after 175 cubic metres are removed?

18.  A lumber dealer has 632,000 bd. ft. (board feet) of white pine on hand. After the sale of 328,582 bd. ft., how many board feet remain?

19.  A carpenter builds a redwood deck for $1,450. His material, labor, and other costs total $735. What is the profit?

20.  A contractor buys 6,000 bd. ft. of oak flooring. One house requires 1,928 bd. ft. and another needs 1,850 bd. ft. How much flooring is left?

21. The balance in a contractor's checking account is $1,176.00. If $321.00 is withdrawn, what is the new balance?

_____

22. Shown at the right is the floor plan of a two-car detached garage. What is the width of the wall space **A** at the front of the garage?

Note: Use this illustration for problems 22–24.

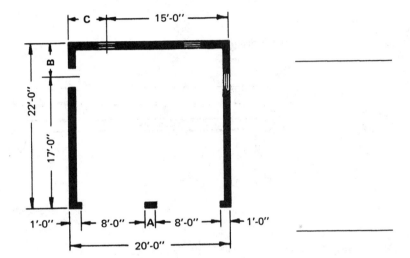

_____

23. What is the distance from the outside of the rear wall of the garage to the center of the side door opening, dimension **B**?

_____

24. What is the length of missing dimension **C** at the rear of the garage?

_____

25. Determine missing dimension **A** in the sketch.

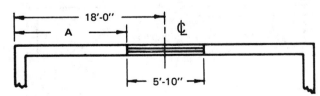

_____

26.   Determine missing dimension **B** in the sketch.

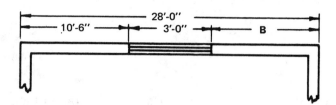

27.   Find missing dimension **X** in the sketch.

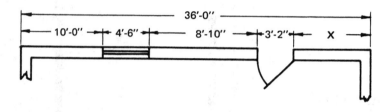

28.   Find missing dimension **D** in the sketch.

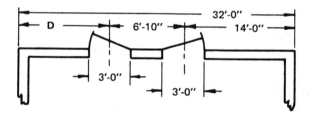

29.   Find the length of missing dimension **E** in the sketch.

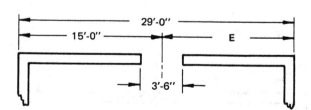

30. A piece of lumber is 18 feet long. After a piece 4 feet 4 inches is sawed off, what length, in feet and inches, is left? _____

31. From an inventory of 1,272 sheets of 1/4-inch paneling, a dealer sold the following number of sheets: 15, 73, 87, 121, 53, 22, and 30. How many sheets of paneling are left? _____

**Note:** When locating the position of a window in a wall during construction, the distance that the opening is placed from one end of the building is usually given on the plans. In the following problems, find the distance from the window to the *other* end of the building.

Note: Use this illustration for problems 32—35.

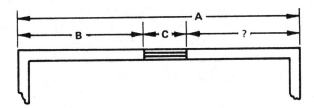

| Prob. No. | Total Wall Length Distance **A** | Width of Window Distance **C** | Distance **B** from End of Building to Window | |
|---|---|---|---|---|
| 32. | 31'-8" | 3'-0" | 6'-7" | _____ |
| 33. | 16'-2" | 3'-6" | 4'-9" | _____ |
| 34. | 30'-6" | 3'-6" | 11'-5" | _____ |
| 35. | 33'-10" | 5'-2" | 13'-1" | _____ |

# Unit 3   MULTIPLICATION OF WHOLE NUMBERS

## REVIEW PROBLEMS

Multiply the following quantities:

| | | | |
|---|---|---|---|
| 1.  16 centimetres<br>x 44 | 4.  352<br>x 75 | 7.  659 yards<br>x 212 | 10.  1,972 board feet<br>x 109 |
| 2.  56 litres<br>x 17 | 5.  1,809 inches<br>x 62 | 8.  350 hours<br>x 521 | 11.  1,257 gallons<br>x 857 |
| 3.  254<br>x 16 | 6.  2 834 kilometres<br>x 170 | 9.  $1,205<br>x 57 | 12.  $16,005<br>x 77 |

13. What is the total length of 25 pieces of door casing if each piece is 7 feet long?                  _____

14. Determine the total wall area of a room if each of the four walls has an area of 357 square feet.  (Do not allow for openings.)                  _____

15. A carpenter places 200 linear feet of joists per hour.  How many feet are placed in 8 hours?                  _____

16. The cost of a new garage is $4,755.  Find the total cost of building 13 garages of this kind.                  _____

17. Find the cost of 28 squares of asphalt shingles at $28.00 per square.  (A square is 100 square feet.)                  _____

18. Allowing 760 shingles per square (100 square feet) when laid 5 inches to the weather, how many shingles are required to cover 24 squares?                  _____

19. The illustration shows the dimensions of a gable roof which is to be shingled.   Determine the number of square feet that the entire roof contains. (Assume that the opposite side has no projection.)

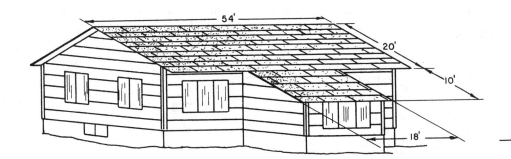

20. A partition is 24'-0'' long by 8'-0'' high. Find the area of one side of the partition. (Hint: The area of a rectangle is found by multiplying the length by the height.) _____

21. If there are 250 cedar shingles in one bundle, how many are there in 258 bundles? _____

22. In making a table, 19 board feet of oak are used, including waste. How much stock is needed to complete an order for 54 such tables? _____

23. A breakfast set requires 43 board feet of lumber. How many board feet are needed for 53 sets that are to be made for a coffee shop? _____

24. Three hundred twenty galvanized nails are required to apply a square of asphalt shingles. How many nails are needed to lay 25 squares? _____

25. How many linear feet of furring are there in 9 bundles, each containing 10 pieces which are 12 feet long? _____

26. How many board feet of lumber are there in the following list of materials: 164 joists — 32 bd. ft. each; 16 girders — 96 bd. ft. each; 58 rafters — 28 bd. ft. each; 14 posts — 44 bd. ft. each; 296 boards — 16 bd. ft. each; 164 studs — 14 bd. ft. each. _____

**Note:** To solve more complex problems, try to divide the problem into a series of smaller, easier steps. Do problems 27 and 28 by following the steps outlined and using the illustrations given.

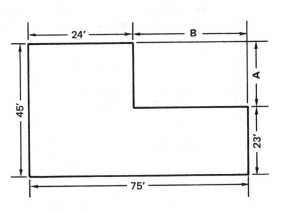

27.  How many square feet of floor space are there in the building shown?

a.  Find missing dimension **A**.

A = _____

b.  Divide the total area into rectangular Sections I and II.
(Other correct divisions are
possible.)  Find the areas of
Sections I and II.  (Area =
length x width)

Section I = _____

Section II = _____

c.  Add the areas of Section I and
II to find the total area.

Total = _____

28.  The walls of the building are 11'-0'' high.  Find the total area of the
outside walls.

a.  Find missing dimensions **A** and **B**.

A = _____

B = _____

b.  Find the area of each wall.  (Area
= length x height)

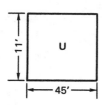
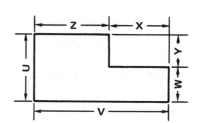

Area of wall **U** = 45' x 11' = _____ .     **U** = _____ sq. ft.
Area of wall **V** = 75' x 11' = _____ .     **V** = _____ sq. ft.
Area of wall **W** = 23' x 11' = _____ .     **W** = _____ sq. ft.
Area of wall **X** = B x 11'  = _____ .     **X** = _____ sq. ft.
Area of wall **Y** = A x 11'  = _____ .     **Y** = _____ sq. ft.
Area of wall **Z** = 24' x 11' = _____ .     **Z** = _____ sq. ft.

c.  Total the individual areas.

_____

Note: Use this illustration for problems 29 and 30.

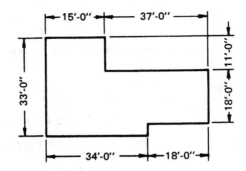

29. Determine the number of square feet of exterior wall area in the building as illustrated. The walls are 14'-0" high. _____

30. How many square feet of floor area does the building contain? (Do not allow for the thickness of the walls.) _____

# Unit 4   DIVISION OF WHOLE NUMBERS

## REVIEW PROBLEMS

Divide the following quantities:

1.  846 inches ÷ 6 =  _____

2.  8,916 pounds ÷ 6 =  _____

3.  15,553 ÷ 2 =  _____

4.  6,921 ÷ 14 =  _____

5.  3 434 centimetres ÷ 34 =  _____

6.  $26,334 ÷ 77 =  _____

7.  4,068 ÷ 142 =  _____

8.  26 104 grams ÷ 26 =  _____

9.  217 124 millimetres ÷ 206 =  _____

10. How many pieces of lumber 36 inches long can be cut from a piece of lumber 216 inches long?  (Do not allow for saw kerfs.)  _____

11. How many hours are needed to lay 902 square feet of subflooring at the rate of 82 square feet per hour?  _____

12. One person can apply 24 square feet of siding per hour.  At the same rate, how long does it take a crew of 6 to put on 11,952 square feet of siding?  _____

13. How many hours does it take to install 3,192 square feet of batt-type insulation at the rate of 152 square feet per hour?  _____

14. How many joists spaced 16 inches o.c. (on center) are required for a floor 52 feet long?

Note:  Add one joist for a starter.

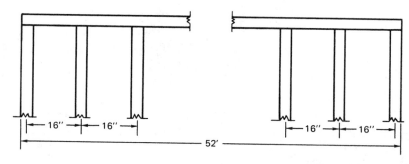

15. Find the number of studs spaced 16 inches o.c. required for a load-carrying partition 64 feet long.
    **Note:** Add one stud for a starter.  _____

16. A common gable roof is 34'-0'' long.  How many rafters spaced 24'' o.c. are required for one side of the roof?
    **Note:** Add one rafter for a starter.  _____

17. The illustration shows a short flight of steps. Dimension **A** is the height or rise of each step. Determine dimension **A**.

Note: Use this illustration for problems 17 and 18.

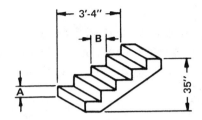

18. Determine the run of each step as indicated by dimension **B** in the illustration. (Hint: Express 3'-4" in inches.)

_____

19. A gable roof is 72'-0" long. Rafters are spaced 24" o.c. How many rafters are needed for both sides of the roof?
    **Note:** Add one rafter for each side of the roof as a starter.

_____

20. The main stairway in a building has 17 risers. If the story height (distance from top of the 1st story floor to the top of the 2nd story floor) is 9'-11", what is the height of each riser?

_____

**Note:** For problems 21–25 allow for only one set of joists.

21. How many floor joists, as illustrated, are required for a building 48'-0" long if the joists are placed 16" o.c.?

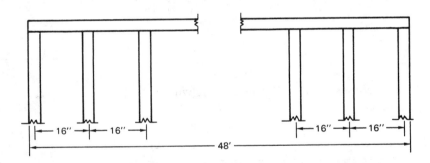

_____

22. How many joists, spaced 16" o.c., are needed for a building that is 36 feet long?

_____

23. How many joists, placed 16" o.c., are required for a building that is 60 feet long?

_____

24. The specifications for a building state that the joists are to be placed 24 inches on center. How many floor joists are required if the building is 116 feet long?

_____

25. A warehouse is 116 feet long. To support the heavy load on the floor, joists are spaced 12 inches o.c. How many joists do the specifications require?

_____

26. A warehouse is 98 feet wide by 164 feet long. Girders run the long way of the building and are spaced 14 feet o.c. How many linear feet of girder are required?

   _____

27. A building is 42 feet wide and 80 feet long. Girders running the length of the building are spaced 6 feet o.c. The outside walls rest on a foundation. Supporting each girder are piers spaced 5 feet o.c. How many piers are needed?

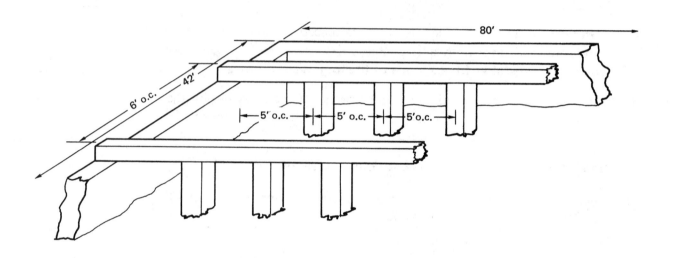

   _____

28. Determine the number of sheets of 4' x 8' plywood subfloor needed to cover the floor area as shown.

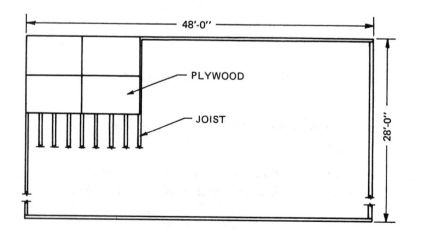

   _____

# Common Fractions

## Unit 5   ADDITION OF COMMON FRACTIONS

### BASIC PRINCIPLES OF ADDITION

A fraction is a part of a whole.  A whole inch may be divided into equal parts in many ways.  For instance, an inch may be divided into eight equal parts.  Each part is known as an eighth of an inch.

EIGHTHS

Should five of these parts be needed in measuring a length, it would be five-eights of an inch and written 5/8''.

The *number of parts* is ⟶ $\dfrac{5}{8}$ ⟵ The *whole* is called the denominator.
called the numerator.

To add fractions, each fraction must first be expressed with a denominator common to all other fractions in the problem.

**Example:**    1/2 + 1/3 + 5/6 + 4/9
        18/36 + 12/36 + 30/36 + 16/36

A common denominator may always be found by multiplying all the denominators of the problem; this often results in an extremely large number.

**Example:**    1/2 + 1/3 + 5/6 + 4/9 ($2 \times 3 \times 6 \times 9 = 324$)

A least common denominator (lowest common multiple) is generally found by inspection.

### REVIEW PROBLEMS

Add the following fractional quantities:

1.  3/4 inch + 5/8 inch          _____

2.  1/8 + 1/3 + 1/6          _____

3.  3/8 gallon + 7/8 gallon          _____

4.  1/4 yard + 1/2 yard + 1/3 yard    _____

5.  5/6 square foot + 3/4 square foot + 3/8 square foot + 1/3 square foot    _____

6.  3 board feet + 7 1/2 board feet + 19 3/4 board feet    _____

7.  2/3 hour + 2 1/4 hours + 4 1/3 hours    _____

8.  What is the total thickness of three boards 5/16 inch, 5/8 inch, and 7/8 inch thick?    _____

9.  What is the total thickness of a table top made of 3/4-inch particle board covered with 1/16-inch laminated plastic?    _____

10.  A desk top is 3/4 inch thick.  It is covered with plate glass 1/4 inch thick.  Find the total thickness.    _____

11.  Find the total thickness of three pieces of plywood, 5/8 inch, 3/8 inch, and 3/4 inch thick.    _____

**Note:**  Use this illustration for problems 12–16. All stock is 3/4″ thick. Top overhang is 1″ on each side. Shelves have 1/4″ dado.  (Refer to the glossary when an unfamiliar term is used.)

12.  Determine the width of the top of the illustrated cabinet.    _____

13.  Give the height of the cabinet.    _____

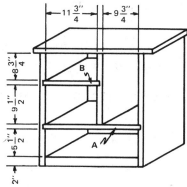

14.  Determine the length of shelf **A**.    _____

15.  Give the length of shelf **B**.    _____

16.  What is the distance from the top of shelf **A** to the top of the cabinet.    _____

17.  A section of the outside wall of a frame building is shown. What is the wall thickness?    _____

18. The insulation board sheathing on the exterior face of the wall in the illustration is covered with patterned cedar shingles. These add 5/8 inch to the thickness. What is the thickness of the wall after it is shingled?

Note: Use this illustration for problems 17–19.

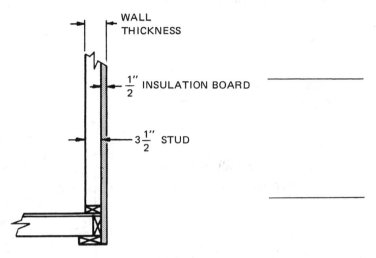

19. After applying 5/8″ cedar panels to the exterior, the interior face of the stud is covered with dry wall 1/2″ thick. What is the total thickness of the wall?

_____

20. Dry wall partitions separate the rooms of a house. If the partition studs are 3 1/2 inches thick, and the dry wall on each face of the stud is 1/2 inch thick, what is the thickness of the partition?

_____

21. How thick is a panel built up of plywood 3/8 inch, 1/4 inch, 5/8 inch, and 3/4 inch thick?

_____

22. The base of a platform is built of lumber 2 5/8 inches thick. It is covered with boards 3/4 inch thick. Find the total thickness.

_____

23. A table top 7/8 inch thick is covered with 1/32-inch veneer. What is the total thickness of 5 of these table tops?

_____

24. A shed door is made of 2 pieces of 1/2-inch and 1 piece of 3/8-inch plywood. Find the total thickness of the door.

_____

25. Find the total width needed for three wall tiles 4 inches by 4 inches if 1/8 inch is allowed between tiles for grouting.

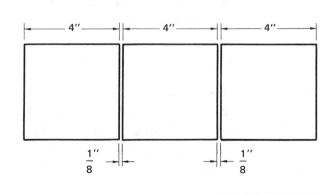

_____

**Note:**   Use this illustration for problems 26–30.   All stock is 3/4″ thick.   Shelves have 1/4″ dado.

26. What is the overall height of the illustrated bookcase?

_____

27. What is the overall length of the shelves of this bookcase?

_____

28. How high from the floor is the top of shelf **A**?

_____

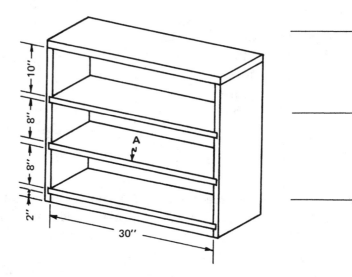

29. Find the length of the top of the bookcase.

_____

30. What is the height of the finished pieces used for sides of the bookcase?

_____

31. A carpenter cuts 3 pieces of stock of different widths from one board. The first is 2 1/4 inches, the second is 3 5/16 inches, and the third is 4 1/4 inches wide.  If 1/8 inch is allowed for each saw cut, find the width of the board used.

_____

32. The foreman on a job gives an apprentice carpenter the job of ripping off a 7/8-inch piece, a 1 1/4-inch piece, a 2 7/16-inch piece, and a 3 3/4-inch piece from a board.  If 1/8 inch is allowed for each saw cut, how much of the board's width is used?

_____

33. What width board is required for the stair treads shown?

_____

Note: Use this illustration for problems 33–35.

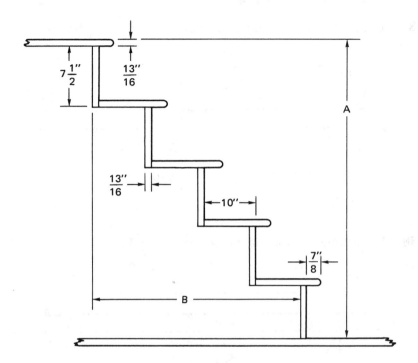

34. What is the vertical distance **A**?

_____

35. What is the horizontal distance **B**?

_____

36. How much is taken from the depth of the drawer shown in the illustration if the groove is 3/8 inch from the bottom and the drawer bottom is 1/4 inch thick?

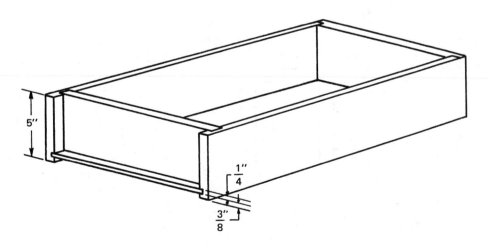

_____

**Note:**  Answers to problems 37—40 should be in simplest form.  For example,
3'-17 1/2″  =  3' + 12″ + 5 1/2″  =  3' + 1' + 5 1/2″  =  4'-5 1/2″

37.   Determine the missing dimension in the diagram below.

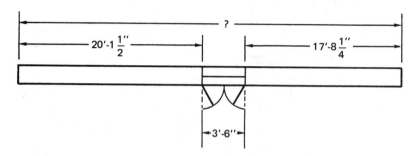

38.   Determine the missing dimension in the diagram below.

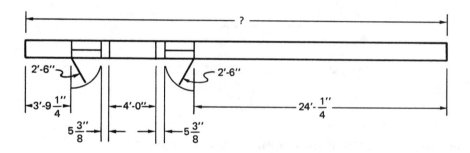

39.   Determine the missing dimension in the diagram below.

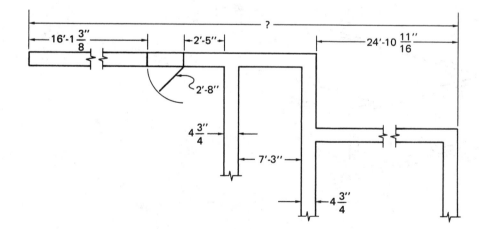

**40.**   What is the height of the chest shown below?

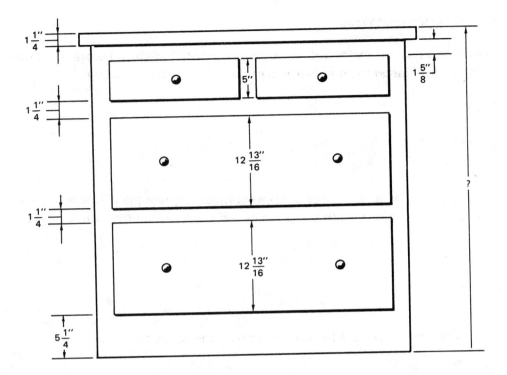

# Unit 6  SUBTRACTION OF COMMON FRACTIONS

## BASIC PRINCIPLES OF SUBTRACTION

As in addition, fractions must be expressed with common denominators before subtracting. Subtract the numerators and express the difference over the common denominator.

**Example:**  Subtract $\frac{1}{2}$ from $\frac{2}{3}$.

$$\frac{2}{3} - \frac{1}{2} = \frac{4}{6} - \frac{3}{6} = \frac{4-3}{6} = \frac{1}{6}$$

To subtract a fraction from a whole number or a mixed number, express the whole number as an improper fraction, with a denominator common to the denominator of the fraction.

**Example:**  Subtract $\frac{2}{3}$ from 5.

$$5 - \frac{2}{3} = \frac{15}{3} - \frac{2}{3}$$

Next, subtract the numerators and express the difference over the common denominator.

**Example:**  $\frac{15-2}{3} = \frac{13}{3} = 4\frac{1}{3}$

## REVIEW PROBLEMS

Subtract the following quantities:

1.  3/8 gallon from 3/4 gallon  _____

2.  1/4 litre from 7/8 litre  _____

3.  1/2 board foot from 7/8 board foot  _____

4.  2/3 hour from 3/4 hour  _____

5.  1/3 from 7/9  _____

6.  7/8 inch from 1 5/8 inches  _____

7.  1/4 metre from 1 1/8 metre  _____

8.  7/32 from 25/32  _____

9.  5/8 inch from 29/32 inch  _____

10.  3/4 inch from 31/32 inch  _____

11. Find the thickness of a rough board 7/8 inch thick after 1/16 inch is planed off one side.   _____

12. A rough board 7/8 inch thick has 1/8 inch taken off by planing on one side. What is its thickness?   _____

13. A counter top is laminated with 1/16-inch plastic. The total thickness is 13/16 inch. What is the thickness of the top?   _____

14. How much must a 7/8-inch board be planed to make it the required thickness of 25/32 inch?   _____

15. The thickness of a piece of three-ply wood is 1/2 inch. It was made by gluing 1/8-inch veneer to both faces of a core. Find the thickness of the core stock.   _____

**Note:** Nails are measured in *penny* or *d* units. A 2 penny nail, abbreviated 2d, is 1 inch long. Lengths of nails are found on a nail chart. Use this chart to solve problem 16.

| Penny (d) Size | Length In Inches | Wire Gauge | | | |
|---|---|---|---|---|---|
| | | Common Nails | Box Nails | Finish Nails | Casing Nails |
| 2 | 1 | 15 | 16 | 17 | 16 |
| 3 | 1 1/4 | 14 | 15 | 16 | 15 |
| 4 | 1 1/2 | 12 | 14 | 15 | 14 |
| 5 | 1 3/4 | 12 | 13 | 14 | 14 |
| 6 | 2 | 11 | 12 | 13 | 13 |
| 7 | 2 1/4 | ... | 12 | 12 | 12 |
| 8 | 2 1/2 | 10 | 11 | 12 | 11 |
| 10 | 3 | 9 | 11 | 11 | 11 |
| 12 | 3 1/4 | 9 | ... | 11 | 10 |
| 16 | 3 1/2 | 8 | ... | 11 | 10 |
| 20 | 4 | 6 | ... | 10 | 9 |

**NAIL CHART**

16. How much longer is a 12d nail than an 8d nail?   _____

17. Find the difference in width of two pieces of hardwood flooring 25/32'' x 2 5/8'' and 25/32'' x 1 5/8''.   _____

**Note:** Answers for all problems should be in simplest form.  For example,
3'-21 1/2" = 3' + 12" + 9 1/2" = 3' + 1' + 9 1/2" = 4'-9 1/2"

18.  As shown in the illustration, the rail for a table is to have a 3/4-inch tenon cut on each end.  If the finished length of the stock is 30 inches, what is the distance between the shoulders?

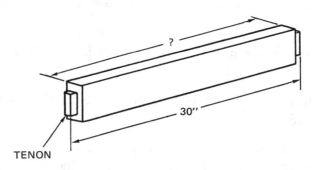

TENON

_____

Note:  Use this illustration for problems 19–23

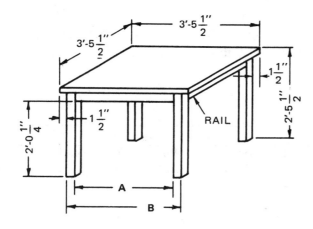

19.  The illustration shows an oblique view of a table.  The table top is 3/4 inch thick.  How long are the table legs?

_____

20.  The legs on this table are 2 1/4 inches square.  What is the length of the rail between the legs?

_____

21.  If the table top is 3/4 inch thick, what is the width of the rail?

_____

22.  Find measurement **B** from the outside face of one table leg to the outside face of the other leg.

_____

23.  How long should the rail be cut if 1 1/4 inches are allowed on each end for the tenons?  (The legs are 2 1/4 inches square.)

_____

24. Turned spindles for a room divider are purchased in standard lengths of 60 inches. The opening in which they are used has a height of 56 7/8 inches. How much must be cut off each spindle? _____

25. By how much does the length of a 3/4" x 1 5/8" rectangular tile exceed its width? _____

26. What is the final thickness of a 3-inch piece of material if 1/8 inch is planed off both faces? _____

**Note:** Use this illustration for problems 27–30. All stock is 3/4" thick. Shelves have dado 3/8" deep.

27. What is the overall length of the shelves in the illustration of the cabinet?

28. The cabinet shown is made entirely of 3/4" stock. What is the inside distance between the sides of this cabinet?

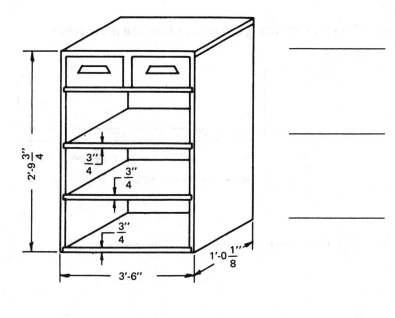

29. The shelves are spaced 11 inches, top-to-top. What is the distance between two shelves?

30. The back of the cabinet is 3/4 inch thick. What is the depth of the shelves if the back sets flush with the back edge of the sides?

31. Four decorative shutters, each 10 inches wide, are installed in the "pass-thru" opening shown in the illustration. How much must be taken off the sides of the shutters to make them fit the opening?

Note: Use this illustration for problems 31 and 32.

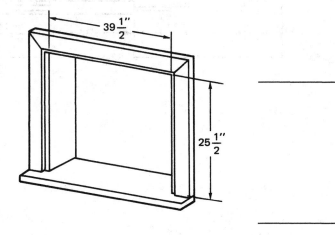

32. Each shutter installed in this opening has a height of 25 7/8 inches. How much must be cut off the top of each one to make it fit?

33. A casing for a beam is constructed as shown in the illustration. What thickness of stock is needed for the members which makes up the sides?

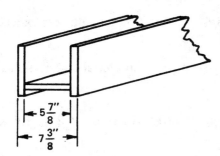

_____

34. Determine missing dimension **A** in the illustration.

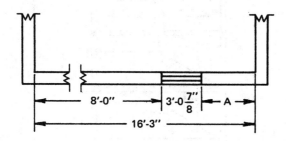

_____

35. Determine missing dimension **B** in the illustration.

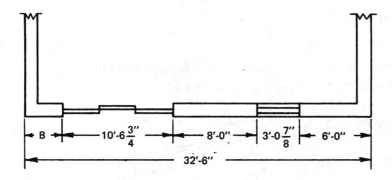

_____

# Unit 7   MULTIPLICATION OF COMMON FRACTIONS

## BASIC PRINCIPLES OF MULTIPLICATION

When multiplying two or more common fractions, multiply the numerators then multiply the denominators and reduce the fraction to lowest terms.

**Example:**   $\dfrac{3}{4} \times \dfrac{2}{5} = \dfrac{3 \times 2}{4 \times 5} = \dfrac{6}{20} = \dfrac{6 \div 2}{20 \div 2} = \dfrac{3}{10}$

When multiplying fractions by whole numbers, change the whole number to a fraction. Proceed as in multiplying common fractions.

**Example:**   $\dfrac{5}{8} \times 3 = \dfrac{5}{8} \times \dfrac{3}{1} = \dfrac{5 \times 3}{8 \times 1} = \dfrac{15}{8} = 1\dfrac{7}{8}$

When multiplying fractions by mixed numbers, change the mixed number to an improper fraction. Proceed as in multiplying common fractions.

**Example:**   $\dfrac{1}{3} \times 3\dfrac{3}{5} = \dfrac{1}{3} \times \dfrac{18}{5} = \dfrac{1 \times 18}{3 \times 5} = \dfrac{18}{15} = 1\dfrac{3}{15} = 1\dfrac{1}{5}$

## REVIEW PROBLEMS

Multiply the following quantities:

1. 5/8 x 3/8 inch _____
2. 7/8 x 4/5 _____
3. 2 3/8 x 4 1/4 gallons _____
4. 5 3/8 x 2 7/8 inches _____
5. 10 1/3 x 3 2/3 _____

6. 7/8 x 5/8 _____
7. 3/4 x 1/8 board foot _____
8. 3/4 x 5/8 _____
9. 7/8 x 1/4 _____
10. 2/3 x 7/8 _____

11. A floor space is covered by 38 boards. Each board has a 3 5/8-inch exposed surface. Find the width of the floor. _____

12. There are 14 risers in the stairs from the basement to the first floor of a house. Find the total height if each riser is 7 1/8 inches high. _____

13. Shingles are laid so that 4 3/4 inches are exposed in each course. How many vertical feet of roof are covered by 28 courses? _____

14. A board 5 3/4 inches wide is cut to 3/4 its original width. Find the new width. _____

15. A beam is 25 1/2 feet long. Find 3/4 of its length. _____

16. The flight of stairs shown has 9 risers.  The height of each riser is 7 5/8 inches. What is the height from floor to floor (total rise)?

17. The run of each step on the flight of stairs is 9 3/8 inches.  What is the total run of the flight?

Note:  Use this illustration for problems 16 and 17.

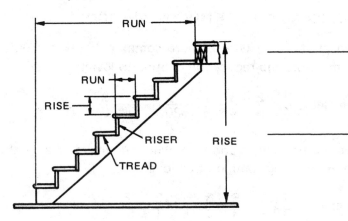

_____

_____

18. A flight of stairs has 14 risers, each 7 3/4 inches high.  What is the total rise?

_____

19. A piece of plywood is 1/2 inch thick.  It is increased by gluing two 3/8-inch pieces to it.  How thick is the finished product?

_____

20. A laborer can place 9 linear feet of sills in 1/4 hour.  At this rate find the time necessary to place 126 linear feet.

_____

21. If 1/4 inch on a drawing represents 1'-0'', how many inches on the drawing represent 18'-0''?

_____

22. Four pieces of wood each 11 3/8 inches long are needed to build a cabinet.  To get all four pieces from one board, how long must that board be?  Allow 1 inch for waste.

_____

23. A board is to be ripped into four strips.  Each one must be 2 3/4 inches wide.  Allowing 7/8 inch for waste in milling the stock, what width board should be used?

_____

24. Find the number of hours required to install 15 roof trusses if it takes 1/3 hour per truss.

_____

25. What is the total thickness, in inches, of 25 table tops if each is 7/8 inch thick?

_____

26. When 1'' x 8'' bevel siding is applied to a building with 6 3/4'' exposed to the weather, what height will be covered with 15 strips? (Allow 8'' for the top strip.)

_____

27. In a flight of stairs there are 13 treads. Each tread has a run of 9 1/4 inches. What is the total run for the flight of stairs? _____

28. What length of 2″ x 4″ material is required to make 6 bench legs each 2 feet 4 1/4 inches long? _____

29. A portion of the rear elevation of a frame dwelling is shown. The clapboards are spaced 4 3/4 inches to the weather. By counting the clapboards in the figure, determine dimensions **A, B, C, D,** and **E**. Express the answers in inches.

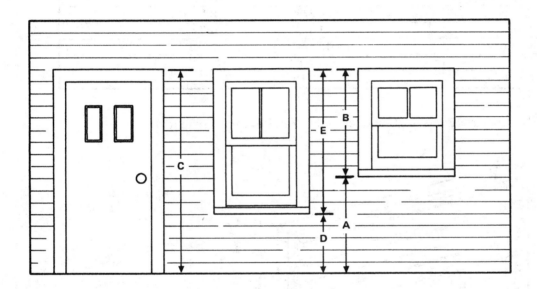

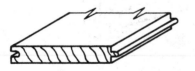

A _____
B _____
C _____
D _____
E _____

**Note:** Matched flooring (Tongue and Groove) has a tongue on one edge as shown. Waste occurs in milling. An original width of 4 inches becomes only 3 1/4 inches on its face when milled. Because of waste, extra stock is needed. If 1″ x 4″ flooring is used, an additional 1/4 of the area to be covered is added. If 6-inch flooring is used, 1/6 of the area is added.

30. The floor of a vacation home has an area of 700 square feet. It is to be covered with 1″ x 4″ T. and G. (Tongue and Groove) flooring. How much additional flooring should be ordered to allow for waste? _____

31. A barn loft with an area of 4,254 square feet requires a matched floor. If 1″ x 6″ T. and G. flooring is used, how much additional stock is needed to allow for waste? _____

32. The first floor of a two-story house has an area of 1,014 square feet. The second floor has an area of 744 square feet. If 1" x 4" T. and G. flooring is used, find the total allowance, in square feet, for waste.   _____

33. What length of each kind of stock is needed to fill the order in the chart below?  (Remember that the number of saw cuts is always one less than the number of pieces needed.)

| | Material | Quantity | Length | Allowance for Each Saw Cut | Total Length Required |
|---|---|---|---|---|---|
| a. | Sculptured Molding | 10 | 8 1/2" | 1/16" | |
| b. | Door Casing | 4 | 30 3/8" | 1/8" | |
| c. | Dowel Rod | 32 | 1 1/8" | 1/16" | |
| d. | Quarter Round | 6 | 15 1/4" | 1/8" | |
| e. | Cove Molding | 9 | 10 7/8" | 1/8" | |
| f. | Screen Molding | 5 | 3 1/16" | 3/32" | |

34. Each rail of the fence shown is 5 1/2 inches wide.  Find the total height of the fence.

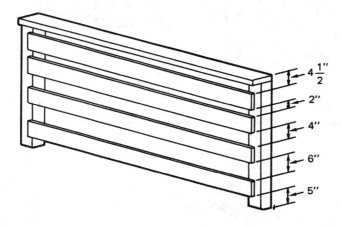

35. An area divider has vertical slats each 5 3/4 inches wide.  Determine the total width of the section of divider in the illustration.

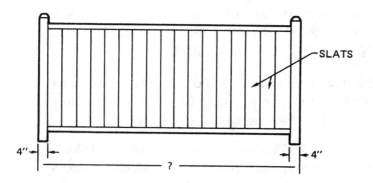

# Unit 8 DIVISION OF COMMON FRACTIONS

## BASIC PRINCIPLES OF DIVISION

The operation of division is very similar to multiplication. When dividing a common fraction, invert the divisor and proceed as in multiplication.

**Example:** Divide $\frac{1}{3}$ by $\frac{1}{2}$.

$$\frac{1}{3} \div \frac{1}{2} = \frac{1}{3} \times \frac{2}{1} = \frac{1 \times 2}{3 \times 1} = \frac{2}{3}$$

When dividing fractions and whole numbers or mixed numbers, change the entire problem to fractions.

**Examples:** $4 \div \frac{1}{3} = \frac{4}{1} \div \frac{1}{3} = \frac{4}{1} \times \frac{3}{1} = \frac{12}{1}$ or 12

$$2\frac{1}{4} \div 1\frac{1}{2} = \frac{9}{4} \div \frac{3}{2} = \frac{9}{4} \times \frac{2}{3} = \frac{9 \times 2}{4 \times 3} = \frac{18}{12}$$ or $1\frac{1}{2}$

Note the difference between multiplication and division:

When multiplying or dividing by 1, there is no change in the answer.

$$12 \times 1 = 12 \qquad\qquad 12 \div 1 = 12$$

When multiplying by a number greater than 1, the product is larger than the original number. When dividing by a number greater than 1, the answer is smaller than the original number.

$$12 \times 3 = 36 \qquad\qquad 12 \div 3 = 4$$
$$\text{(Increased)} \qquad\qquad \text{(Decreased)}$$

When multiplying or dividing by a number less than 1, the results are reversed, i.e., the product is decreased and the quotient is increased.

$$12 \times \frac{1}{2} = 6 \qquad\qquad 12 \div \frac{1}{2} = 24$$
$$\text{(Decreased)} \qquad\qquad \text{(Increased)}$$

## REVIEW PROBLEMS

Divide the following quantities:

1. 3/4 ÷ 3 _____

2. 3/8 mile ÷ 3 _____

3. 4/9 yard ÷ 1/9 _____

4. 3/4 ÷ 7/8 _____

5.  5/8 gallon ÷ 3/4  _____

8.  3 3/8 ounces ÷ 6  _____

6.  3 inches ÷ 3/8 inch  _____

9.  5 3/8 inch ÷ 4 1/8  _____

7.  7/8 ÷ 3/8  _____

10.  22 1/8 board feet ÷ 1/2  _____

11.  How many pieces of 7/8-inch lumber are there in a stack 35 inches high?  _____

12.  Determine the number of pieces of 3/8-inch plywood necessary to make a panel 1 1/2 inches thick.  _____

13.  A line is drawn so that 1/2 cm represents 1 m.  How many metres are represented by a line 4 centimetres long?  _____

14.  Determine the number of pieces of 3/8-inch plywood in a stack 30 inches high.  _____

15.  A stack of table tops is 60 inches high.  Find the number of table tops if each is 3/4 inch thick.  _____

16.  How many shelf boards 4 1/2 feet long can be cut from an 18-foot board?  _____

17.  How many pieces of floor covering 3 5/8 inches wide are needed to cover 36 1/4 inches of floor width?  _____

**Note:**  Before dividing denominate numbers, express both terms using the same units.  Example: $5'\text{-}4\frac{1''}{2} \div 1\frac{1''}{2} = 64\frac{1''}{2} \div 1\frac{1''}{2} = \frac{129}{2} \div \frac{3}{2} = \frac{129}{2} \times \frac{2}{3} = 41$

18.  How many boards 4 5/8 inches wide does it take to cover a floor 18 1/2 feet wide?  _____

19.  In the illustration, the front elevation of a lumber rack is shown.  What is the width of space **A** if the three spaces are of equal width?

Note:  Use this illustration for problems 19 and 20.

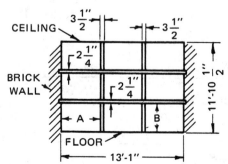

20.  What is the height of space **B** if all spaces are of equal height?  _____

21. A board 10 inches wide is ripped into strips 2 5/16 inches wide. If a total of 3/4 inch is allowed for milling the strips, how many strips are there?   _____

22. How many boards, each 2 feet 4 inches long, can be cut from a length of stock 14 feet long? Make no allowance for saw cuts.   _____

23. How many supporting columns 7 feet 4 inches long can be cut from 6 pieces, each 22 feet long?   _____

24. How many pieces of 1/2-inch plywood are there in a stack 3 feet 6 inches in height?   _____

25. If 1/4 inch represents 1 foot on a drawing, how many feet are represented by 10 1/8 inches?   _____

26. How many 7 1/2-inch risers are there in a flight of stairs 7 feet 6 inches high?   _____

27. A carpenter divides a dimension stick 5 feet 3 inches long (shown below) into 12 equal spaces. How long is each space?

_____

28. A dimension stick is 6 feet 2 inches long. It is laid off in spaces 4 5/8 inches long. How many spaces are there?   _____

29. How many 6 1/8-inch spaces are there on a dimension stick 8 feet 2 inches long?   _____

30. If a dimension stick is 6 feet 4 inches long, and is laid off in spaces 4 3/4 inches long, how many spaces are there?   _____

31. When 1/8 inch represents 1 foot on a drawing, how many feet are represented by 6 1/2 inches?   _____

32. A carpenter lays 10 1/2 squares of wood shingles in 4 1/2 days. Find the average number of squares laid in one day.   _____

33. A flight of stairs is 7 feet 11 7/8 inches high. Each riser is 7 3/8 inches high. How many risers are there?   _____

34. A garage is covered with 6-inch siding. Each board lays 5 1/4 inches. The height of the wall is 8'-3 3/4". How many boards are needed?   _____

35. Three additional holes are to be drilled through the piece of stock shown.  Locate the centers by finding distance **A**.

_____

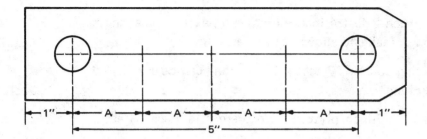

36. A small building is 16 feet 6 1/4 inches wide and 20 feet long.  How many 20-foot flooring boards are needed to cover the floor if each board lays 3 1/4 inches?

_____

37. Ten-inch siding covers 9 1/8 inches.  How many boards are required for a wall that is 16 feet 2 inches high?  (Count fractional boards as whole boards.)

_____

38. The landing of a stairway is 4 feet 11 inches high, and each riser is 7 3/8 inches high.  How many risers does it take to reach the first landing?

_____

39. The distance from floor to floor on a two-story house is 9 feet 8 inches.  How many risers are there in the stairway if the plans call for a 7 1/4-inch rise?

_____

40. A bookcase is 5 feet 8 1/4 inches high.  The five shelves, each 3/4 inch thick, are equally spaced.  Determine the distance between the shelves if the top of the lowest shelf is 4 inches from the floor.

_____

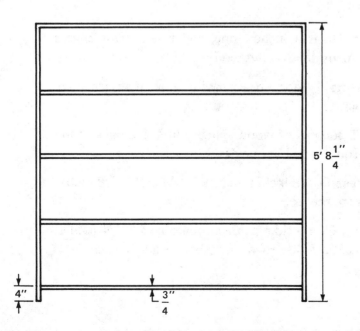

**Note:** Before a carpenter can start to build stairs, several measurements must be determined. These measurements include:

Total rise
Total run
Rise of each step
Width of each tread

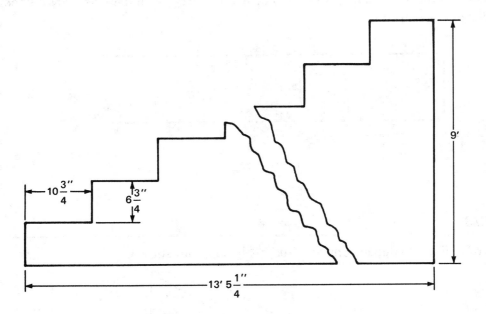

41. How many risers are needed for the set of stairs illustrated?  _____

42. How many treads are needed for a set of stairs with a total run of 13 feet 5 1/4 inches when each tread is 10 3/4 inches wide?  _____

# Unit 9  PRACTICE WITH COMMON FRACTIONS

## BASIC PRINCIPLES OF CARPENTER'S RULES

Carpenter's rules are divided into inches by lines marked with the number of inches.  The next shorter lines indicate the half inches; still shorter lines indicate the quarter inches, eighth inches and sixteenth inches.  Some fine quality rules indicate the thirty second inches.  Note these markings on the following illustration.

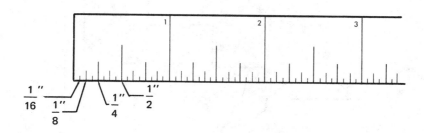

## REVIEW PROBLEMS

Using a carpenter's rule, find the dimensions required of the lap joint shown.

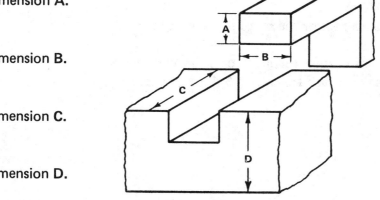

1.  Find dimension A.
_____

2.  Find dimension B.
_____

3.  Find dimension C.
_____

4.  Find dimension D.
_____

5.  How many 1/100ths of a metre are there in 3 metres?
_____

6.  How many 1/8ths of an inch are there in 1/4 inch?
_____

7.  How many 1/8ths of an inch are there in 3/4 inch?
_____

8.  How many 1/8ths of an inch are there in 1 1/2 inches?
_____

9.  How many 1/10ths of a centimetre are there in 4 centimetres?
_____

10.  How many 1/16ths of an inch are there in 3/8 inch?
_____

11. How many 1/16ths of an inch are there in 5/8 inch? _____

12. How many 1/16ths of an inch are there in 3/4 inch? _____

13. How many 1/16ths of an inch equal 1/4 inch? _____

14. How many 1/16ths of an inch equal 7/8 inch? _____

15. How many 1/8ths of an inch equal 3/4 inch? _____

16. What is the length of a piece of stock 1/8 inch longer than 1/4 inch? _____

17. What is the length of a piece of stock 1/16 inch longer than 1/2 inch? _____

18. What is the length of a piece of stock 1/4 inch longer than 5/8 inch? _____

19. What is the length of a piece of stock 1/16 inch longer than 2 3/8 inch? _____

20. Give the rule reading for each of the lettered dimensions.

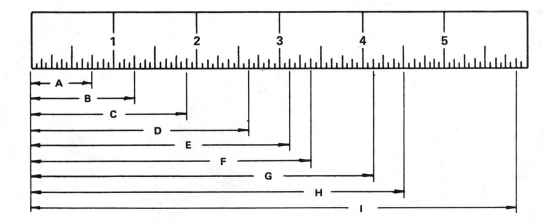

A _____

B _____

C _____

D _____

E _____

F _____

G _____

H _____

I _____

21. Give the rule reading for each of the lettered dimensions.

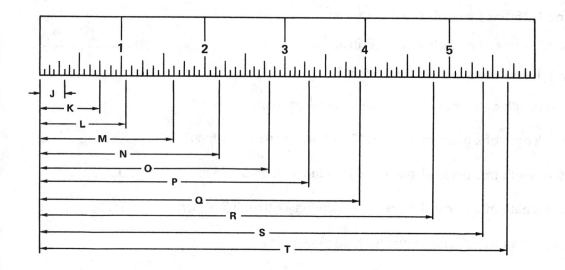

J _____

K _____

L _____

M _____

N _____

O _____

P _____

Q _____

R _____

S _____

T _____

Note:  Use this illustration for problems 22 and 23.

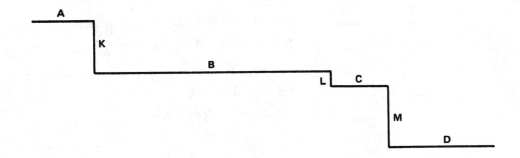

22.  Measure lines **A, B, C,** and **D** in the illustration and indicate the total length in inches.                                                          _____

23.  Measure lines **K, L,** and **M,** and give the total length in inches.                                                          _____

Note:  Use this illustration for problems 24 and 25.

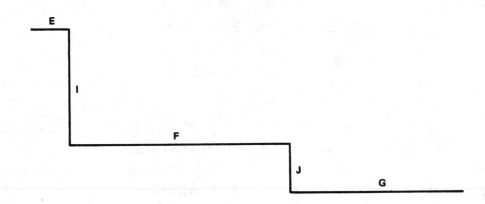

24.  Measure lines **E, F,** and **G,** and give the total length in centimetres.                                 _____

25.  Measure lines **I** and **J,** and give the total length in centimetres.                                 _____

Solve the following problems.  Express all answers in simplest form.

26.  3/8 + 5/8 _____

27.  1/4 + 1/8 _____

28.  1/2 gallon + 3/8 gallon _____

29.  5/16" + 7/16" _____

30.  3/16" + 1/4" _____

31.  7/8 pound + 1/16 pound _____

32.  9/16" + 1/4" _____

33.  9/16 + 1 5/8 + 1/4 _____

34.  2 3/8" + 1 5/16" + 3/4" _____

35.  3 1/8" + 1/2" + 7/16" _____

36.  1 11/16" + 5/8" + 2" _____

37.  1" – 3/8" _____

38.  3/4" – 1/8" _____

39.  1/2 board foot – 3/16 board foot _____

40.  1 1/4" – 1/16" _____

41.  1 1/2" – 5/16" _____

42.  2 3/4 – 5/8 _____

43.  3 5/8" – 7/16" _____

44.  13/16" – 3/8" + 1 5/8" _____

45.  1 1/16 + 7/16 – 1/2 _____

46.  2 7/8" – 1 5/16" + 1/4" _____

47.  3" + 2 1/4" – 7/8" _____

48.  2 7/16 + 3/8 – 15/16 _____

49.  8 x 10 1/2 cm _____

50.  1 3/4 g x 1/2 _____

51.  3 5/8" x 1/4 _____

52.  2 1/8" x 4 1/2 _____

53.  5 1/8" x 1 1/4 _____

54.  11 1/2 ÷ 4 _____

55.  8 3/4" ÷ 1/2" _____

56.  10 3/4" ÷ 2 _____

57.  2 x 3 1/2 m ÷ 3/4 m _____

58.  5 x 6 1/4" ÷ 3 3/4" _____

59.  4 x 1/8" x 1 1/2" _____

60.  2 1/4 x 1/8 ÷ 2 _____

61.  1'-3 1/2" + 4'-6 3/4" – 3'-2 1/4" _____

62.  5'-1 1/4" – 5 7/8" + 2'-3 1/8" _____

63.  1'-3" x 1/4 _____

64.  4 1/2 m x 3 _____

65.  5'-3 1/2" x 1/2 _____

66.  2'-4" x 5 1/4" _____

67.  2'-3 1/2" ÷ 1 1/2" _____

68.  3 1/2' ÷ 4" _____

69.  6 3/4' ÷ 7/8" _____

70.  7 1/2' ÷ 1 1/2" _____

71.  What fractional part of a foot is 4 inches? _____

72.  What fractional part of a foot is 3 1/2 inches? _____

73.  Find 1/2 of 10 3/4 inches? _____

74.  Find 1/4 of 3' 4". _____

75.  Find 1/2 of 5' 8 1/2". _____

# Decimal Fractions

## Unit 10   ADDITION OF DECIMAL FRACTIONS

### BASIC PRINCIPLES OF ADDITION

Decimal fractions are fractions with denominators which are powers of ten.

**Example:**   $\dfrac{3}{10}, \dfrac{7}{100}, \dfrac{13}{1000}$

When these fractions are written in decimal form, they are called decimal fractions or decimals.

**Example:**   0.3, 0.07, 0.013

To add decimals, arrange the numbers with the decimal points directly under each other.  Locate the decimal point for the sum, then add.

**Example:**

```
    0.3
    4.7
    2.31
 +  0.08
    7.39
```

### REVIEW PROBLEMS

| 1. | 2. | 3. | 4. | 5. |
|---|---|---|---|---|
| 1.56 | 5.329 | 47.362 | 18.053 | 15.763 |
| 0.05 | 0.036 | 89.347 | 0.637 | 4.800 |
| + 0.69 | + 1.56 | + 47.365 | + 193.008 | + 9.005 |

6.  Add 0.325 m and 0.05 m.   _____

7.  Add 0.857 inch and 0.643 inch.   _____

8.  Add 4.152 cm, 31.648 cm, 6.52 cm, and 2.48 cm.   _____

9.  Add 0.0652 inch, 0.3148 inch, 0.5624 inch, and 0.6876 inch.   _____

10.  Add $33.81, $25.69, $13.72, $47.67, and $54.60.   _____

11.  A table top 1.857 inches thick is covered with laminated plastic 0.0625 inch thick.  What is the total thickness of the top?   _____

12. A piece of plywood has an inner core 0.325 inch thick. It is covered by two pieces, each 0.275 inch thick. What is the total thickness?

_____

13. Find the thickness of the plywood block from the information given in the illustration.

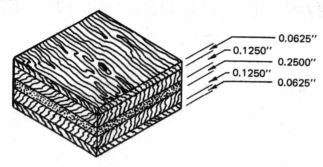

0.0625"
0.1250"
0.2500"
0.1250"
0.0625"

_____

14. Costs for completing six small buildings are $13,152.46; $14,037.92; $14,987.78; $13,490.12; $15,048.61; and $15,409.73. What is the total cost?

_____

15. A contractor pays $1,250.45 for frame lumber, $650.34 for sash and trim, and $156.20 for flooring. What is the total cost?

_____

16. A homeowner buys the following materials: lumber, $31.65; hardware, $4.20; paint, $16.85; and glazing compound $2.75. What is the total cost?

_____

17. A homeowner pays a carpenter $48.00 for labor, $96.48 for shingles, and $6.72 for nails. What is the total cost of labor and material?

_____

18. What is the total cost of the following order: crown molding, $18.20; base shoe, $8.76; chair rail, $26.15; quarter round, $12.76?

_____

19. A carpenter pays $120.50 for shingles, $8.25 for nails, and $112.00 for labor. What is the total amount paid?

_____

20. A carpenter receives $460.25 for one job and $575.40 for another job. How much is received for both jobs?

_____

21. A contractor submits a bill of $1,906.42 for framing material and trim, $65.50 for hardware, $462.00 for masonry, $170.35 for painting, and $850.67 for labor. What is the total amount for the items listed?

_____

22. Records show that the material on a job cost $1,289.45, the labor $678.92, and the overhead $69.45. The contractor wants to realize a profit of $128.90. Find the total cost to the owner.

_____

23. A carpenter working on three jobs collects $62.36 for the first job, $43.42 for the second, and $21.92 for the third. How much is collected for the three jobs?

_____

24. The bracket shown must be constructed for a specific job. Find the minimum width of material needed to make a pattern for this bracket.

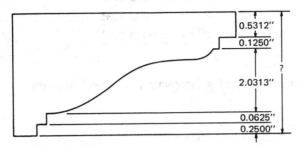

0.5312"
0.1250"
?
2.0313"
0.0625"
0.2500"

_____

25. In bidding for a job, a contractor lists the following items: material, $1,257.45; labor, $928.75; trucking the equipment to the job, $16.20; over-head, $39.75; and profit, $235.50. What is the total bid to be submitted? _____

26. Find the total cost of the following hardware bill:

| | |
|---|---|
| Nails | $16.40 |
| Locks | 29.44 |
| Drawer pulls | 5.40 |
| Cupboard catches | 3.00 |
| Hinges, 2" x 2" butts | 4.75 |
| Bolts, carriage | 4.34 |
| Screening | 21.89 |
| Door bumper | 1.75 |
| Coat and hat hooks | 3.60 |
| Handles | 1.65 |
| Elbow catches | 0.50 |
| Window lifts | 2.50 |
| Window locks | 3.90 |

_____

27. A carpenter installs several new doors and hardware. Total the following bill for materials:

| | |
|---|---|
| Hollow core doors | $81.20 |
| Lock sets | 27.30 |
| Hinges | 12.20 |

_____

28. A cabinet shop receives estimates of materials for the construction of built-in cabinets. What is the total?

| | |
|---|---|
| Lumber | $320.00 |
| Hinges | 30.96 |
| Magnetic catches | 24.00 |
| Contact cement | 8.87 |
| Drawer pulls | 32.60 |

_____

29. Find the total number of board feet in the following lumber list:

     Sill                    160.75 board feet
     Studs                   426.67 board feet
     Plate                   106.64 board feet
                                                                    _____

30. Determine the total height, in inches, of a stairway with the following dimensions:

     Floor to first landing      15.5 inches
     Landing to second floor   78.75 inches
                                                                    _____

31. What is the total cost for the following hardware?

     Hinges                  $9.75
     Locks                    7.60
                                                                    _____

32. Use the following charges to find the total cost for laying a finish floor:

     Flooring                $65.20
     Finish                    8.35
     Labor                    46.30
                                                                    _____

33. Determine the total cost of the following materials:

     Crown molding           $15.60
     Bed molding              37.80
     Chair rail                7.70
     Corner mold               4.50
     Shoe base                17.50
                                                                    _____

34. From the drawing, determine the total length of the irregular template.

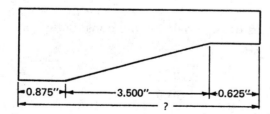
                                                                    _____

35. A contractor receives bills of $123.61; $97.43; $1,040.70; $507.38; and $1,231.65 for materials. What is the total amount of the bills?  _____

36. An expense record is as follows: material, $1,349.20; labor, $897.65; salaried help, $375.00; and overhead expenses, $275.45. What are the total expenses?  _____

37. The diagram shows a custom-made adapter shaft. A carpenter is to construct a cabinet to house this adapter shaft for reasons of safety. If the adapter shaft requires at least 1 inch of clearance on each end, what is the minimum interior length, in inches, of the cabinet?

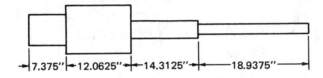

_____

# Unit 11  SUBTRACTION OF DECIMAL FRACTIONS

## BASIC PRINCIPLES OF SUBTRACTION

In order to subtract decimal fractions, arrange the numbers with the decimal points directly under each other: Locate the decimal point for the difference, then subtract.

**Example:**

$$
\begin{array}{r}
392.47 \\
-\ \ 41.29 \\
\hline
351.18
\end{array}
$$

When subtracting decimal fractions, it is sometimes necessary to add zeros.

**Example:**

$$
\begin{array}{r}
84.65 \\
-\ 63.523
\end{array}
\qquad
\begin{array}{r}
84.650 \longleftarrow \text{ZERO ADDED} \\
-\ 63.523
\end{array}
$$

## REVIEW PROBLEMS

1.  $4,326.75
    $-\ 2,138.47$
    _____

2.  2.703 km
    $-1.937$ km
    _____

3.  $1.23 – $0.68
    _____

4.  4 327.5 g – 13.6 g
    _____

5.  0.326 – 0.187
    _____

6.  Subtract $68.37 from $91.41.
    _____

7.  Subtract 1.35 litres from 2.65 litres.
    _____

8.  Subtract 0.365 from 0.635.
    _____

9.  Subtract 1.87 km from 4 km.
    _____

10. Subtract 143.5 miles from 156.6 miles.
    _____

11. The diameter of a 10d nail is 0.148 inch.  The diameter of a 6d nail is 0.113 inch.  How much greater is the diameter of a 10d nail than that of a 6d nail?
    _____

12. The length of a lot is 159.5 feet and the width is 38.5 feet.  By how much does the length of the lot exceed its width?
    _____

13. Weights of two pieces of tempered hardboard are 0.97 pound and 1.21 pounds per square foot. Find the difference in weight per square foot. _____

14. A contract is accepted for $25,050.00. The total cost of material and labor is $22,709.79. How much is the profit on the contract? _____

15. The wrought iron ornamental grille shown is to be installed in an opening of 94.625 inches. How much must be cut off?

96.75"

_____

16. A contractor accepts a job for $575.00. Bills for materials and labor are $113.52, $287.61, and $78.92. What is the profit? _____

17. A carpenter receives $1,400.00 for building a deck. Materials cost $671.54. How much is received for labor and profit? _____

18. A carpenter pays $1,647.82 for material and receives $2,567.00 for a job. Assuming no other costs, what is the cost of labor? _____

19. An exterior door costs $135.00 glazed with single-strength glass or $199.80 with insulating glass. What is the difference in price? _____

20. A carpenter receives $421.25 for laying finish floor. The nails and flooring cost $257.60. How much is received for labor and profit? _____

21. A carpenter estimates the cost of purchasing and installing a cabinet at $70.95. If the price of the cabinet is $51.45 delivered at the job complete, what is the cost of the labor? _____

22. An estimate of $1,250.75 is submitted for a job. The actual material cost is $495.42, and the labor and overhead come to $596.36. What profit does the contractor expect on the job? _____

23. The labor for milling some stock totals 34.75 hours and is charged to three different jobs. To the first job, 24.5 hours are charged, and to the second job, 3.25 hours are charged. How much time is charged to the third job? _____

24. In a mill, four jobs total 146 hours. Of the 146 hours, 26.75 are charged to job No. 1; 46.25 to job No. 2; and 39.25 to job No. 3. The carpenter forgets to record the hours on job No. 4. How many hours are charged to job No. 4?                                                                    _____

25. Find dimension **A** in the pattern illustrated.

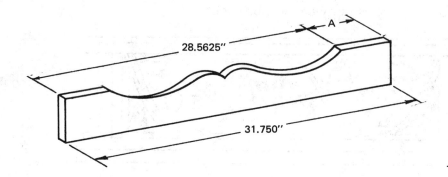

26. A wall mirror costs $25.40. With bevel plate glass it costs $47.65. What is the additional cost of the bevel plate?                                            _____

27. A carpenter pays $26.40 for stock but charges the customer $42.60. What is the profit?                                                                                              _____

28. A contractor receives $231.77 for doing a piece of work. Expenses are $32.46, $35.88, and $129.68. How much profit is realized on this job?                                                                                                              _____

29. The labor for a job amounts to $560.40. The total cost to the owner is $932.70. What is the cost of material?                                                        _____

30. A carpenter and apprentice work a total of 76.4 hours on a job. If the carpenter's time is 38.5 hours, how much time does the apprentice work?                                                                                                      _____

31. An estimate of $181.75 is submitted for a job. If $44.25 is for labor, what is the cost of material?                                                                  _____

32. If 14.75 squares of shingles are delivered to a job and 12.25 squares are used, how many squares are left over?                                                _____

33. What is the length of the Lally column in this illustration, including top and bottom plates?

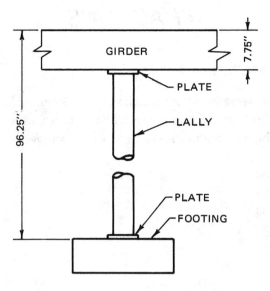

34. A cabinet mill estimate for interior finish is as follows:

| | |
|---|---|
| Doors | $162.78 |
| Windows | 97.22 |
| Casement sash | 26.38 |
| Interior trim | 84.73 |
| Shelving stock | 27.69 |
| Drawer stock | 18.52 |

When the job is complete, the contractor receives a credit of $14.27 on the doors, $3.25 on the sash, and $6.29 on the interior trim. Additional materials are ordered costing $6.26 and $3.75. What is the net amount of the bill?

35. A contractor's bank balance is $3,426.38. A deposit (money put into the bank) of $567.72 is made and the following payroll checks are drawn:

| | |
|---|---|
| Carpenters | $326.16 |
| Bricklayers | 183.98 |
| Plasterers | 462.87 |
| Laborers | 89.28 |
| Painters | 231.29 |

What is the net (final) bank balance?

36. The contract for the lumber on a certain job amounts to $689.67. The millwork bid is $538.73. Additional lumber and millwork costs are $63.51 and $34.50. Credits of $10.23 and $13.48 are received for the return of unused materials. What is the total net cost of the lumber and millwork?

# Unit 12  MULTIPLICATION OF DECIMAL FRACTIONS

## BASIC PRINCIPLES OF MULTIPLICATION

Decimal fractions are multiplied the same as whole numbers. The number of decimal places to the right of the decimal point in both quantities being multiplied are added. This total is the number of places to the right that the decimal point is placed in the answer.

Examples:

```
    43.21              .004
 X   5.42           X .021
    8642               004
   17284               008
   21605            .000084
  234.1982
```

## REVIEW PROBLEMS

1. Multiply 1.75 m by 2.95. _____

2. Multiply 6.29 inches by 4.99. _____

3. Multiply 2.357 g by 2.57. _____

4. Multiply 3.57 yards by 5.52. _____

5. Multiply 87.69 by 25.30. _____

6. 0.0389 x 0.0056 _____

7. 0.947 x 0.004 feet _____

8. 76.9 x 38.2 mm _____

9. 4.76 x 9.83 _____

10. 1 000 x 0.365 mL _____

11. A builder drives 18.4 miles round trip to and from a house under construction. If 117 days are spent working on this house, what is the total mileage? _____

12. A worker must reglaze 27 windows for a hen house. Each window contains six panes of glass. If each pane requires eight points, how many glazing points are needed? _____

13. Find the weight of 534 square feet of 1/8-inch asphalt tile if the average weight per square foot is 1.2 pounds. _____

14. What is the weight of 752 square feet of 1/8-inch vinyl asbestos tile if the average weight per square foot is 1.29 pounds? _____

15. The weight of 1/4-inch plywood is 0.79 pounds per square foot. What is the weight of 1,125 square feet?  _____

16. Determine the weight of a 4' x 5' pane of 3/16-inch plate glass if the weight per square foot is 1.7 pounds.  _____

17. If one square foot of wall requires 6.25 firebricks, find the approximate number needed for 33 square feet.  _____

18. A board is divided into 8 equal segments. If each segment is 12.250 inches long, what is the total length, in inches, of the board?

12.250''

?

_____

19. Laying fiberglass roofing on an irregular roof is estimated to take 2.6 hours per square. How long will it take to lay 14.4 squares?  _____

20. The material needed for the floor of a deck is estimated to be 81.9 board feet, at a cost of $0.65 per board foot. What is the total cost of the material?  _____

21. The fitting of a casement sash of a certain size takes an average of 0.375 hour. How many hours will it take to fit 26 sashes?  _____

22. A carpenter can lay 100 square feet of deadening felt over a subfloor in 0.4 of an hour. How much time does it take to lay 55.25 squares? (One square = 100 square feet.)  _____

23. Determine the total weight of 75 cartons of hardwood block flooring if one carton weighs 26.5 pounds.  _____

24. The illustration shows the floor plan of a reception room in a motel. The floor is to be covered with prefinished block flooring. If it takes 1.3 blocks to cover 1 square foot of floor, find the number of blocks needed to cover the entire floor. (Note: Area of a rectangle equals length times width.)

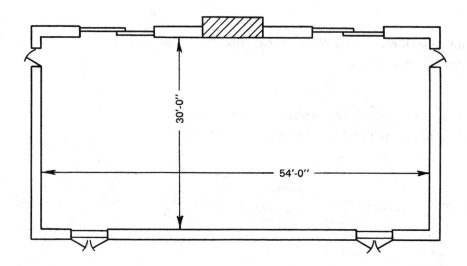

_____

25. Finish flooring is estimated to take 3.2 hours per 100 square feet (one square). At that rate, how much time does it take to lay a floor 15.3 squares in area?

_____

# Unit 13   DIVISION OF DECIMAL FRACTIONS

## BASIC PRINCIPLES OF DIVISION

The numbers in a division problem all have names.

**Example:**   20 ÷ 0.4 = 50

$$\begin{array}{r} 50 \longleftarrow \text{quotient} \\ \text{divisor} \longrightarrow 0.4\overline{)\,20} \longleftarrow \text{dividend} \end{array}$$

Move the decimal point in the divisor to the extreme right, making the divisor a whole number.

**Example:**   $0.4\overline{)20}$

Move the decimal point in the dividend to the right the same number of places. Place zeros after the dividend if necessary.

**Example:**   $4.\overline{)20.0}$     1 zero added

Place a decimal point for the quotient directly above the new decimal point in the dividend.

**Example:**   $4.\overline{)200.}$

Divide as with whole numbers.

**Examples:**

$$\begin{array}{r} 50. \\ 4\overline{)200.} \\ \underline{20}\phantom{0} \\ 0 \\ \underline{0} \end{array} \qquad \begin{array}{r} 3.471 \\ .21.\overline{)72.891} \\ \underline{63}\phantom{000} \\ 98\phantom{00} \\ \underline{84}\phantom{00} \\ 149\phantom{0} \\ \underline{147}\phantom{0} \\ 21 \\ \underline{21} \end{array} \qquad \begin{array}{r} 7\ 00. \\ .04.\overline{)28.00.} \\ \underline{28}\phantom{000} \\ 00\phantom{00} \\ \underline{00}\phantom{00} \\ 00\phantom{0} \\ \underline{00}\phantom{0} \end{array}$$

## REVIEW PROBLEMS

Note:  Round answers to number of decimal places specified.

1.  Divide 7.4 inches by 6.  (3 places) _____

2.  Divide 16 m by 0.57.  (2 places) _____

3.  Divide 8.57 by 1.52.  (3 places) _____

4.  Divide 5.652 by 0.0652.  (3 places) _____

5.  Divide $95.63 by 3.5.  (2 places) _____

6.  Divide 64 pounds by 0.32. _____

7.  Divide 64 by 1.6. _____

8.  Divide 0.208 litre by 0.13. _____

9.  Divide $4 by $0.08. _____

10. Divide 0.4428 by 246.0. (4 places) _____

11. Divide 6.4 by 1. _____

12. Divide 64 g by 0.1 g. _____

13. Divide 64 by 0.001. _____

14. Is the result in division always a number which is smaller than the quantity being divided? _____

15. In the figure shown each of 21 courses of clapboard is equally exposed to the weather. How many inches on each clapboard are *laid to the weather?*

16. How many pieces of plywood, each 0.375 inch thick, are there in a stack 20.25 inches high? _____

17. A stack of table tops each 1.875 inches thick is 3 feet 9 inches high. How many table tops are there in the stack? (**Note:** Express feet as inches.) _____

18. A worker's paycheck for a job is $90. This is two and one-half times as much as the helper's check. How much does the helper receive? (Hint: Express two and one-half as a decimal fraction.) _____

19. A mechanic purchases a package of shims. Each is 0.072 inches thick. The package measures 4.896 inches. How many shims does the package contain? _____

20. A painter spends 12 hours (720 minutes) staining 1,500 square feet on a shingled building. How many minutes will it take to stain 100 square feet? _____

21. Vinyl to finish a kitchen floor costs $398.52 for labor and material. What is the cost per square foot if the area of the floor is 108 square feet? _____

22. How many pieces of 2.25-inch face flooring are required for a closet 24.75 inches deep? _____

23. The total run of a set of stairs is 103.5 inches. Each tread is 8.625 inches. How many treads are there? _____

24. A platform 141.75 inches wide is to be covered with 5.25-inch face sheathing. How many pieces of sheathing are required? _____

25. The distance from the bottom of the wall sheathing to the bottom of the frieze is 99.75 inches.  If the siding is laid 4.75 inches to the weather, how many courses are required?

    _____

26. Determine distance **A** in order to locate the centers for drilling two additional holes in the pattern shown.

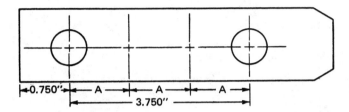

    _____

# Unit 14 EXPRESSING COMMON FRACTIONS AND MIXED NUMBERS AS DECIMALS

## BASIC PRINCIPLES OF CONVERSION

To change a fraction to a decimal, divide the numerator by the denominator.

**Example:** Change 3/8 to a decimal.

```
      .375
  8)3.000
    2 4
      60
      56
      40
      40
```

Those fractions yielding a repeating decimal should be rounded off to the desired number of places.

**Example:** 1/3

```
      .3333
  3)1.0000
```

1/3 = .33333333. . .

Round off to .33 or .333

In order to express decimals as fractions, drop the decimal point and write the given number as the numerator. Write the denominator as a power of ten (10, 100, 1,000, etc.) using as many zeros as there are decimal places in the decimal number.

**Examples:** $.31 = \frac{31}{100}$; $.025 = \frac{25}{1000}$; $5.4 = 5\frac{4}{10}$ or $\frac{54}{10}$

## REVIEW PROBLEMS

Express the common fractions as decimal fractions in problems 1 through 5, and decimal fractions as common fractions in 6 through 10. Refer to the table of decimal equivalents for fractional parts of an inch.

1. 3/8 _____

2. 3/4 _____

3. 1/4 _____

4. 27/32 _____

5. 7/8 _____

6. 0.25 _____

7. 0.78125 _____

8. 0.6875 _____

9. 0.53125 _____

10. 0.546875 _____

11. Approximately 1 1/4 pounds of nails are needed for each 100 square feet of 1" x 10" subflooring. Write this weight in decimal form.   _____

12. The actual width of a pine board is 7 1/4 inches. Write the width in decimal form.   _____

13. The approximate thickness of a piece of asphalt tile is 1/8 inch. Write this thickness in decimal form.   _____

14. The thickness of a piece of oak flooring is 25/32 inch. Express this thickness in decimal form.   _____

15. The actual thickness of a piece of textured panel is 7/16 inch. What is the thickness, written as a decimal fraction?   _____

16. The weight of 1/4-inch plywood is 0.73 pound per square foot. Write this weight in common fraction form.   _____

17. A piece of vinyl is 0.127 inch thick. Express this thickness as a common fraction.   _____

18. A piece of lining felt is 0.041 inch thick. What is the thickness written as a common fraction?   _____

19. A piece of plywood is 0.631 inch thick. What is its thickness in common fraction form?   _____

20. Find the approximate thickness, in common fraction form, of a piece of siding 0.625 inch thick.   _____

**Note:** The answers to problems 21–30 should be expressed correct to the nearest thousandth (the third decimal place).

Note: Use this illustration for
problems 21 and 22

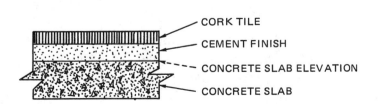

CORK TILE

CEMENT FINISH

CONCRETE SLAB ELEVATION

CONCRETE SLAB

21.  The elevation at the top of the concrete floor shown is 127.7 feet. The cork tile is 0.5 inch thick and the cement finish is 1.25 inches thick. What is the elevation at the top of the cork tile floor?

    a.    0.5 inch  =                      _____ feet

    b.    1.25 inch  =                  _____ feet

        (Elevation of concrete floor  =  127.7 feet)

    c.    Total elevation  =                  _____ feet

22.  Determine the elevation for the cork tile floor shown, if the tile is 1/48 foot and the concrete floor elevation is 141.23 feet.

    a.    Express 1/48 foot in decimal form.            _____

    b.    Find the total elevation.                   _____

**Note:**  In some problems, one dimension is given as a common fraction and another dimension is given as a decimal fraction.  In such cases it is helpful to write both terms in decimal form, or both terms in fraction form.  For example:

        2 1/2 feet  +  0.25 foot  =  2.5 feet  +  0.25 foot  =  2.75 feet

                              or

           2 1/2 feet  +  0.25 foot  =  2 1/2 feet  +  1/4 foot  =
                   2 2/4 feet  +  1/4 foot  =  2 3/4 feet

Notice that 2.75 feet  =  2 3/4 feet  =  2' 9''

23.  A floor has a concrete slab with an elevation of 113.45 feet.  It is to have two cement finishes with a total thickness of 3/4 inch.  What is the elevation of the finished floor?

    a.    Express 3/4 inch as a decimal fraction.           _____ inch

    b.    Express the answer to part (a) as a fraction of a foot in decimal form.                         _____ foot

    (The elevation of the concrete slab is 113.45 feet.)

    c.    Find the total elevation of the finished floor.    _____ feet

          Note:  Use this illustration for problems 24–26

24.  The elevation at the top of a girder as shown below is 171.46 feet. What is the elevation, in feet, of the top of the 7/8-inch wood subfloor?

    a.    Express 7/8 inch as a fraction of a foot in decimal form.           _____ foot

    b.    Find the elevation of the subfloor.          _____ feet

25. The top of a girder is 156.92 feet in elevation. What is the elevation, in feet, of the top of the 25/32-inch wood floor?

   a. Express fractional dimensions in decimal form.
      7/8 inch =  _____ inch
      25/32 inch =  _____ inch

   b. Express these dimensions as feet in decimal form.
      7/8 inch =  _____ foot
      25/32 inch =  _____ foot

   c. Find the total elevation in feet.  _____ feet

**Note:** To solve problems 26—30, use a method similar to problem 25.

26. What is the elevation, in feet, at the top of the terrazzo floor in the figure, if the elevation at the top of the girder is 127.95 feet?  _____

27. In setting form work to pour the finish on a sidewalk, a carpenter is asked to raise the form 3/4 inch above an elevation of 141.35 feet. What is the new elevation?  _____

Note: Use this illustration for problems 28—30.

28. At what elevation must the illustrated concrete setting bed be laid to obtain the required elevation at the top of the tile floor?  _____

29. What is the elevation for the top of the 25/32-inch wood floor in the figure?  _____

30. What is the elevation at the top of the girder in the figure?  _____

31. Express 0.625 inch as 8ths of an inch. _____

32. Express 0.18 inch to the nearest 16th of an inch. _____

33. Express 2.76 inch to the nearest 4th of an inch. _____

34. Express 0.58333 foot to the nearest inch. _____

35. Express 0.17 foot to the nearest inch. _____

36. Express 7.76 feet as feet and inches to the nearest inch. _____

37. Express 21.28 feet as feet and inches to the nearest 8th of an inch. _____

38. Express 7.8 feet as feet and inches to the nearest 8th of an inch. _____

39. Express 13.41 feet as feet and inches to the nearest 16th of an inch. _____

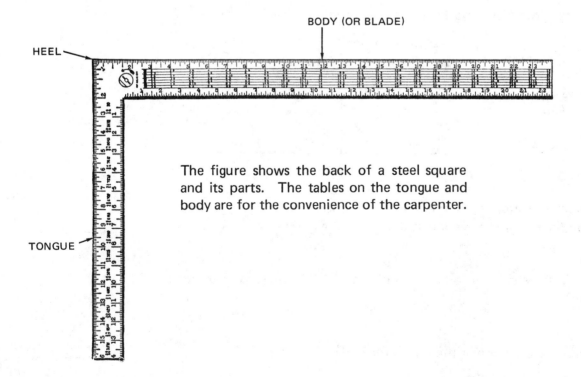

BODY (OR BLADE)

HEEL

TONGUE

The figure shows the back of a steel square and its parts.  The tables on the tongue and body are for the convenience of the carpenter.

**Note:** The figure below shows a section of the brace tables given on the tongue of a steel square. The numbers in the middle give the lengths of the diagonals of squares. For example, notice the circled numbers under the 8-inch mark. This means that if a square has sides with length 48 inches (or 48 feet), the diagonal of that square is 67.88 inches (or 67.88 feet).

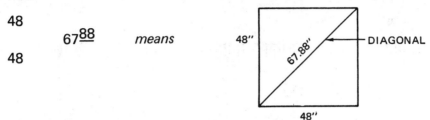

Note: Use this illustration for problems 40–43

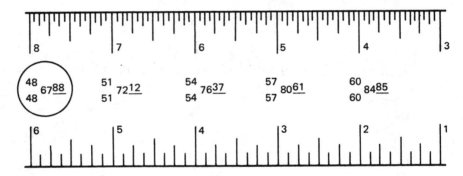

40. How many feet and inches, correct to the nearest 8th inch, are contained in 67.88 inches? _____

41. Notice the numbers under the 6-inch mark. A square with sides of 54 feet has a diagonal of 76.37 feet. Express the length of the diagonal in feet and inches, correct to the nearest 4th inch.

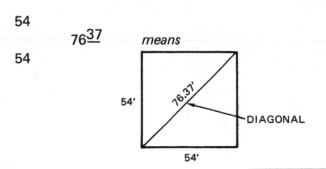

42. Under the 5-inch mark, find the length for the diagonal of a 57-inch square. Express this length in feet and inches, correct to the nearest 64th inch. _____

43. Find the length of the diagonal of a 60-inch square, in feet and inches, correct to the nearest 32nd inch. _____

44. The illustrations below represent segments of a carpenter's rule. Working directly from the rule, at each arrow express the reading as its decimal equivalent.

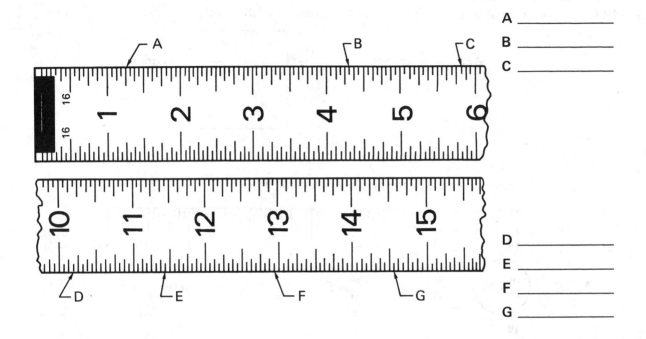

A _____

B _____

C _____

D _____

E _____

F _____

G _____

# Percent and Percentage

## Unit 15 SIMPLE PERCENT AND PERCENTAGE

### BASIC PRINCIPLES OF PERCENT

Percent means *each hundred,* just as miles per hour means miles each hour. "Cent" refers to hundred, as in *cent*ury or *cent*ipede. Forty miles per hour can be expressed as the fraction 40/1; 25% can be expressed as 25/100. Percent may be expressed as a decimal by first changing the number to a fraction and then to a decimal.

**Example:** $37\% = \dfrac{37}{100} = 0.37$

The % sign may be considered a magnet which attracts the decimal point two places to the right. Removing the % sign moves the decimal point two places to the left.

**Example:** 37.% = 0.37

When solving percentage problems, first change the percent to a decimal, and then treat "of" as "times."

**Examples:**  Find 3% of $75.

3% of $75 = 0.03 × $75 = $2.25

Find 150% of 400 lb.

150% of 400 lb = 1.50 × 400 lb = 600 lb

A basic formula for percentage problems is:

Part = Percent × Whole

$P = \% \times W$

The following diagram may help in using this formula.

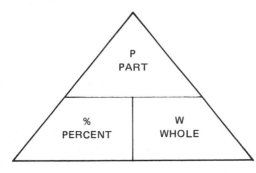

To use this diagram, cover the section you are solving for to determine the formula to use. For example, when determining percent, cover the percent section of the diagram; the fractional parts $P$ (part) and $W$ (whole) remain. Therefore, the formula to use to find percent is:

$$\% = \frac{P}{W}$$

**Examples:**    What percent of 28 is 7?

$$\% = \frac{P}{W}$$

$$\% = \frac{7}{28} = 0.25 = 25.\%$$

$$\% = 25$$

3 is 25% of what number?

$$W = \frac{P}{\%}$$

$$W = \frac{3}{0.25} = 12$$

$$W = 12$$

## REVIEW PROBLEMS

1.  Use the formula *Percentage = Base x Rate* to find the following:

    a. P = 493; R = 85%; B = ?        _____

    b. P = ?; R = 55%; B = 125 m        _____

    c. P = $588; R = ?; B = $840        _____

    d. P = ?; R = 22%; B = 416.8        _____

    e. P = 240; R = ?; B = 192        _____

2.  In the figure shown, what percent of board width is allowed for matching?

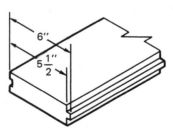

_____

**Note:** To determine the amounts of material needed for various jobs, it is common practice to allow a percentage for waste in cutting.

3. It is estimated that 2,550 board feet of roof boards are needed. If an additional 20% is allowed for waste, find the total amount required.

    _____

4. How many square feet of floor felt should be ordered for a job when the area of the floor is 1,540 square feet and 15% of the area is allowed for waste?

    _____

5. The total area to be covered with insulation board is 1,250 square feet. How much should be ordered if 8% of the area is added for waste?

    _____

6. It is estimated that about 65 pounds of flashing material for drip caps are needed for a certain job. How much should be provided if 20% of the estimate is added to allow for waste?

    _____

7. In making mortar, lime is used in an amount equal to 12% of the cement. How much lime is necessary if 995 pounds of cement are used?

    _____

8. A floor of 385 square feet is to be covered with subflooring laid diagonally. To allow for waste, 25% of the area is added when subflooring is laid diagonally. How many square feet of subflooring should be ordered?

    _____

9. It is estimated that 10,050 facebricks are required for a certain job. How many more bricks are required if 14% of the estimate is added for waste?

    _____

10. A carpenter's helper estimates a need for 280 board feet of flooring but forgets to allow for waste. The carpenter adds 25% to this amount. What is the total number of board feet in the carpenter's order?

    _____

11. Not allowing for waste, a carpenter estimates that 728 board feet of matched flooring are needed for a particular job. If 20% is added for waste and matching, what is the waste allowance for this job?

    _____

12. A carpenter must order subflooring for 2 floors of the same size. Without allowing for waste, the carpenter's apprentice estimates that each floor requires 1,080 feet. If 25% is allowed for waste, how many board feet of subflooring are ordered?

    _____

13. In constructing a building, 2% of the total cost is allowed for excavation. Find the cost of the house if the excavation costs $674.

    _____

14. The carpentry and millwork on a new structure amounts to $6,756. A contractor figures 68% of this amount as the cost of material and the remaining 32% as labor?

   a. What is the material cost?

    _____

   b. What is the labor cost?

    _____

15. A contractor receives $1,800 profit on a job. If this is 12% of the total charge, what does the owner pay?

_____

16. A general contractor estimates a nonresidential building to cost $25,860. If the work of excavating and grading is 3% of this amount; concrete work, 20%; and carpentry, 9%, what is the amount estimated for each of these three items?

_____

_____

_____

17. The total cost of a house is $86,000.00. What percent is spent on carpentry labor and material, if the total for these costs is $42,275.50? (Express answer to nearest tenth percent.)

_____

18. On a nonresidential building, 6% of the total job cost is carpentry labor. If the total cost is $15,000, what is the cost for carpentry labor?

_____

19. The carpentry labor and material on a building is estimated to be $48,500. Of this amount, 47% is for labor and 53% is for material.

a. What is the allowance for labor?

_____

b. What is the allowance for material?

_____

20. What percent of the total area of the piece of stock shown in the figure do the holes represent? Note: Area of a rectangle = length x width. Total area of holes = 2 sq. in.

_____

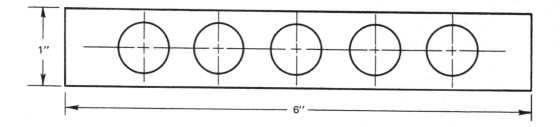

21. The grading on a $45,500.00 house comes to $787.50. What percent of the total cost is this? (Express answer to nearest hundredth percent.)

_____

22. A contractor purchases 17,500 board feet of rough lumber. If there is waste of 18% in milling, how many board feet of lumber are actually received?

_____

23. A certain grade of lumber is supposed to be 75% *clear,* or free from knots and other defects. How much clear lumber should a carpenter expect to find in a load of 2,200 board feet?

_____

24. A dealer buys an item at a cost of $6,600 and sells it at a profit equal to 12% of the cost. Find the amount of profit and the selling price. (Note: Selling price = Cost + Profit.)

_____

_____

25. A contractor pays workers $68.80 per day for labor. If the contractor makes a 10% profit, find the contractor's profit and the charge to the customer. (Note: Charge to customer = Contractor's cost + Contractor's profit.)

_____

_____

26. How much should a contractor bid for a job if costs total $2,300 and a profit of 12% is desired? (Note: Amount of bid = Cost + Profit.)

_____

27. A carpenter charges a customer $862.50 for a job. This includes a profit of $112.50. What is the carpenter's rate of profit?

_____

28. What is the bid on a job with estimated costs of $3,464 if the profit made is 11% of the cost?

_____

29. The estimated cost of a small house totals $51,375. The contractor adds 8% for profit. What is the exact amount of the bid?

_____

30. A contracting company submits a bid for an apartment house. The estimated costs are $181,693.21. To this, add 3% of the estimated cost for overhead, and 9% of the estimated cost plus the overhead for profit. What is the final bid?

_____

31. To an estimate of $43,568.84 a contractor adds 2% for incidentals, and 8% of the estimate and incidentals for profit. What is the total bid?

_____

32. A general contractor's bid shows the following net costs and profits (percent of costs) added by subcontractors:

| | | | |
|---|---|---|---|
| Plumbing | $4,528.10 | + | 7% |
| Masonry | $6,278.43 | + | 11% |
| Electrical work | $1,829.71 | + | 9% |
| Heating | $2,372.17 | + | 6% |

In addition to these costs, the general contractor estimates costs of $61,264.38, and adds a further 12% of the total for profit. What is the general contractor's total bid?

_____

**33.** What percent of the lot illustrated is taken up by the house and driveway? (Note:  Area of a rectangle  =  length x width.)

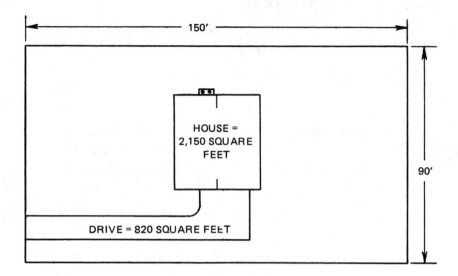

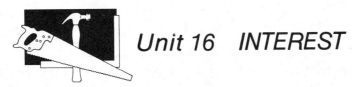

# Unit 16  INTEREST

## BASIC PRINCIPLES OF INTEREST

When a person borrows money, a fee for the use of that money is usually charged by the person lending the money. An example of this is the finance charge that is added for buying items on credit. An amount, in addition to the cost of the item, is charged the customer; this is a form of interest. Banks also pay interest to their depositors for the privilege of using the depositors' money. A carpenter or developer often borrows money to build a house. Interest must be paid on the amount borrowed until the house is sold and the loan can be repaid. Interest is usually expressed as a percent of the principal.

Study the following definitions that relate to computing interest:

*Interest* is a payment for the use of money; it is expressed as a percent of the principal.
*Principal* is the amount of money invested or borrowed.
*Rate of interest* is the percent at which the interest is calculated.
*Amount* includes principal plus previously earned interest.
*Compound interest* is interest based on the principal plus previously earned interest.

## FORMULAS

1. *Simple Interest = Principal x Rate x Time* in years. In symbols, this is *I = PRT.*

2. *Amount = Principal + Interest*

## Examples:

1. Find the simple interest and amount on $800 at 12% after 2 years.

   *Interest = Principal x Rate x Time* (in years)

   *Interest* = $800 x 12% x 2
      = $800 x 0.12 x 2
      = $192

   *Amount = Principal + Interest*

   Amount = $800 + $192
      = $992

2. The annual rate of interest on $1,200 is 14%. Find the interest after a period of 3 months. (Note: 3 months = 3/12 year = 1/4 year)

   Interest = Principal x Rate x Time

   Interest = $1,200 x 0.14 x 1/4
      = $42

## REVIEW PROBLEMS

Determine the simple interest and amount under each of the following conditions.

| | Principal | Rate of Interest | Time | Interest | Amount |
|---|---|---|---|---|---|
| 1. | $    300 | 9% | 4 years | _____ | _____ |
| 2. | $ 3,600 | 10% | 6 years | _____ | _____ |
| 3. | $    640 | 11% | ½ year | _____ | _____ |
| 4. | $    960 | 13% | 1¼ years | _____ | _____ |
| 5. | $ 4,280 | 12% | 3 months | _____ | _____ |
| 6. | $ 6,800 | 14% | 1 year and 6 months | _____ | _____ |
| 7. | $20,000 | 15½% | 1 year | _____ | _____ |
| 8. | $ 1,080 | 9½% | 1½ years | _____ | _____ |

9.  A family pays 13% interest on a $45,000.00 mortgage for the construc-
tion of a house.  What is the yearly interest paid on the mortgage?  _____

10.  A carpenter deposits $192.80 in a savings bank.  If the bank pays 5 1/4%
interest per year, payable annually, what is the total amount in the
account at the end of one year?  _____

11.  A bank loans $4,250.00 to a contractor at a rate of 13% per year.  What
is the yearly interest payment?  _____

12.  A contractor receives $450 interest annually on a mortgage of $3,600.
What is the rate of interest?  _____

13.  A contractor borrows $30,000 on a building loan at 12 1/2% annual
(yearly) interest.  What does the loan money cost for 1 year?  _____

14.  A builder develops a tract of 20 houses at a cost of $46,000 each.  If
the mortgage is placed on one-half of the total cost of the 20 houses,
at 11 1/2% interest, what interest is paid each year?  _____

15.  Two business partners have a bank balance of $7,942.00.  After the
first year their interest brings their balance to $9,053.85.  What rate of
interest is earned?  _____

16.  A contractor borrows the sum of $9,973.00 and pays the bank 14 1/2%
interest yearly.  How much interest is paid for 1 year?  _____

17.  Find the interest for 1 year, at 13 1/2%, on a lumber list of $3,896.75.  _____

**Note:** In certain cases interest is figured monthly, quarterly, or semiannually; the contract must clearly state which. The simple interest formula, $I = PRT$, is used, but $T$ represents the *number of interest periods,* instead of the number of years. (*Interest = Principal x Rate x Time* in number of interest periods.) If interest is figured monthly, the interest period is 1 month.

**Example:** A bond of $3,200.00 carries an interest rate of 3% for one month. Find the interest for 3 months.

Interest = *Principal x Rate x Time* (number of interest periods)
Interest = $3,200 x 0.03 x 3      Interest = $288 Ans.

18. A contractor secures a bond of $2,500.00 at a rate of 3% for one month. If this bond is held for four months, how much does the bond cost for that period?   _____

19. A contractor gives a bank a note for $7,550.00 at a rate of 3% for one month. How much interest is charged for 3 months?   _____

20. A hardware company adds 10% to bills that are not paid within a 30-day period. A carpenter is unable to pay a bill of $70.80 until after 30 days. What is the interest charged on the bill?   _____

21. A lumber yard levies a 12% interest charge on money owed them after 30 days. How much interest does a contractor pay on a $1,876 bill which is paid in 90 days?   _____

22. A builder makes a $1,200 profit on the sale of a new home. The profit is deposited in a bank at 6% interest, compounded quarterly. What is the bank balance at the end of 6 months if no deposits and no withdrawals are made during that time? Use the steps outlined below to solve this problem.

**Note:** *Compound interest* is computed on the original principal *plus previously earned interest.* Interest compounded *quarterly* is added to the account every 3 months (1/4 year); there are 4 interest periods each year. The interest rate for a period equals the annual *interest rate* divided by the number of interest periods in 1 year. To compute compound interest, follow these steps:

a. How many interest periods are there in 1 year?   _____
b. For how many interest periods is the money left in the bank?   _____
c. What is the interest rate for 1 period?   _____
d. Find the interest for the first period. Principal x Rate for 1 period = Interest for 1 period   _____
e. Find the amount at the end of the first period. Principal + Interest = Amount   _____
f. Find the interest for the second period. Amount at end of 1st period x Rate for 1 period = Interest for 2nd period   _____
g. Find the amount at the end of the second period. Amount at end of 1st period + Interest for 2nd period = Amount at end of 2nd period   _____

23.  A contractor has $3,755 in a 6% interest-drawing account.  If interest
     is paid every six months and no withdrawals are made, how much
     money is in the bank at the end of 3 years?  (Hint:  Use the steps
     outlined in problem 22.)

                                                                    _____

# Unit 17   DISCOUNTS

## BASIC PRINCIPLES OF DISCOUNT

*Discount rate* is the percent by which a price is marked down. *Discount* is the amount by which the price has been marked down. The *list price* multiplied by the *discount rate* equals the *discount*:

$$L \times R = D$$

The *list price* minus the *discount* equals the *net price* (sale price).

**Example:**   A hammer marked $12.50 is being sold at a 10% discount.  Find the sale price.

$$\begin{array}{ccc} L & \times & R = & D \\ \$12.50 & \times & 0.10 = & \$1.25 \end{array}$$

$$\begin{array}{ccc} L & - & D = & N \\ \$12.50 & - & \$1.25 = & \$11.25 \end{array}$$

Sometimes a double discount is given, which is a discount on the sale price.

**Example:**   A radial arm saw lists for $400.  It is on sale marked down 10%.  An additional 2% discount is given for paying cash.  If a buyer pays cash, what is the cost?

$400 \times 0.10 = \$40$

$400 - \$40 = \$360$

$360 \times 0.02 = \$7.20$

$360 - \$7.20 = \$352.80$

## REVIEW PROBLEMS

1. List price is $5,670; discount is 12% of this price.  Find the cost.  _____

2. Retail price is $45; discount is 25% of retail price.  Find the cost.  _____

3. List price is $5,780; discount is 12% of the list price; cash discount is 2%.  Find the net cost.  _____

4. Retail price is $58; discount is 20% of retail price; cash discount is 3%.  Find the net cost.  _____

5. Retail price is $36; discount is 25% of retail price; cash discount is 2%.  Find the net cost.  _____

6. List price is $73; discount is 10%; cash discount is 2%.  Find the net cost.  _____

7. A bill is $948.50; discount is 15%; cash discount is 2%.  Find the net cost.  _____

8. An item is on sale at a 25% discount. The sale price is what percent of the list price?

   _____

9. A customer buying through a wholesale outlet pays 80% of the retail price. What percent of the retail price is saved?

   _____

10. Net cost is $28.80 after a discount of 20% is allowed off retail price. Find the retail price. (Hint: The net cost is 100% - 20%, or 80% of the retail price. To find the retail price, divide the net price by the 80%.)

   _____

11. A discount of 2% is allowed for cash. If the amount of cash paid is $1,225.00, find the original amount of the bill.

   _____

12. Cost is $15.86 after taking 35% off catalog price. Find the catalog price. $\overline{165}$

   _____

13. An amount paid is $154 after 12% discount is allowed. Find the original amount of the bill.

   _____

14. Net cost is $5,188.80 after taking 8% off list price. Find the list price.

   _____

15. Net cost is $28.80 after a discount of 15% is allowed off retail price. Find the retail price.

   _____

16. If 2% is allowed for cash, and the amount of cash paid is $1,025.00, find the original amount of the bill.

   _____

17. Net cost is $5,831; discount on selling price is 15%; cash discount is 2%. Find the selling price.

   _____

18. Net cost is $26.46; discount is 25% of retail price; cash discount is 2%. Find the retail price.

   _____

19. Net cost is $630.63; discount on retail price is 10%; cash discount is 2%. Find the retail price.

   _____

20. If cost is $15.86 after taking 20% off the catalog price, find the catalog price.

   _____

21. When the amount paid is $154 after a 15% discount is allowed, determine the original amount of the bill.

   _____

22. List price is $48.92; net cost is $44.03. Find (a) the amount of discount, and (b) the rate of discount. (Express answer to nearest whole percent.)

   _____

   _____

23. List price is $792.47; net cost after cash discount is $776.62. Find (a) the discount, and (b) the rate of discount. (Express answer to nearest whole percent.)

   _____

   _____

24. If a bill is paid within 10 days, the customer is charged a net total of $27.60. If the bill is paid after 10 days, the gross of $28.45 is charged. What is the rate of discount for prompt payment? (Express the answer to the nearest whole percent.)    _____

25. The list price of a steel square in a catalog is $7.70, subject to a 25% discount. What is the net price of the square?    _____

**Note:** It is customary in many lines of business to have a list price on which certain discounts are given. This enables the manufacturer or firm to print prices in their catalogs that will be usable even though the prices of the materials or goods fluctuate. The actual price is changed by varying the discount.

26. A contractor orders material that costs $1,926.90, less 2% if paid within 30 days. How much is due if the bill is paid 10 days after ordering?    _____

27. A jointer plane is listed at $20.35. A carpenter purchases it at list price less 30%. What is the net cost?    _____

28. A carpenter purchases a miter box which lists at $95.76, less 33 1/3%. How much does the box cost? (Hint: Use 33 1/3% = 1/3 to simplify the work.)    _____

29. A contractor receives a large bill for hardware and is given discounts of 15% and 5% off. What is the net cost for the hardware, the price quotation of which is $781.60?    _____

**Note:** In addition to the single discount allowed on some commodities or building materials, business practice often permits the giving of a second one. After the first discount (which is always the largest) is taken off, the second discount is figured on the remainder. In trade parlance the double discount is known as 10 and 2 off, 12 and 4 off, or whatever is relevant.

30. A hardware bill totals $345.69 with discounts of 5% and 3%. What is the net cost of the material?    _____

31. The list price of a table saw is $450 less 20% and 10%. What is the actual cost to a carpenter who can obtain these discounts?    _____

32. A contractor purchases oak lumber for $1,400.00, less 2% discount; yellow pine for $596.00, less 2%; white pine for $896.50, less 1%. What is the total amount of the bill?    _____

33. A contractor receives the following bill and pays it within 30 days, thus receiving an extra discount of 2%. Find the total net cost.

| | |
|---|---|
| Oak treads and risers | $51.60, less 2% |
| Newel posts | 36.75, less 1% |
| Balusters | 64.00, less 2% |

_____

34. A catalog quotation on a bill for millwork amounts to $1,241.50. The firm allows a discount of 2% and 1%. How much does a carpenter have to pay for this material?

_____

35. A tool chest containing 12 items is listed for $95. Discounts of 15% and 10% are allowed. If the customer pays cash within 30 days, an additional discount of 5% is granted. What is the cost of the set if a carpenter takes advantage of all the discounts offered?

_____

36. Window screens are priced at $4.13 each. How much must a carpenter pay for 12 screens with wholesale discounts of 5% and 2%?

_____

37. A contractor gets a job to furnish picnic benches. He can purchase the benches ready-made for $26.25 each, with discounts of 2% and 1%. It will cost the contractor $25.75 to make each picnic bench. (a) Which is cheaper? (b) By how much?

_____

_____

38. The dealer's price to a contractor on a bill for materials is 12 1/2% off list price, with an additional 2% discount for cash within 30 days. (a) Find net price on a $1,975 order. (b) Find the total discount.

_____

_____

39. A contractor gets the following bill from a mill:

| Six window units | $59.01 each, less 1% |
| One corner cabinet | $54.98, less 2% |
| Two 3' 0'' x 6' 8'' | |
| colonial door units | $58.65 each, less 2% |

If an additional 2% cash discount is allowed, what must the contractor pay for the stock?

_____

40. A hardware firm sells nails at a discount of 28% of the list price. What is the cost of 540 pounds of 8d nails that are listed at 31 cents per pound?

_____

41. What is the total cost of 1,870 pounds of 16d nails at 35¢ per pound, list price, and 118 dozen bolts at a list price of 72¢ per dozen? The discount on the nails is 13% and 17% on the bolts.

_____

42. Nails are listed at $21.00 per fifty-pound box; shelf standards at $1.15 each; shelf brackets at $0.54 each; closet rods at $0.96 each; and aluminum track at $0.90 each. What is the cost for the following bill?

| 600 pounds of nails | discount 12% |
| 20 shelf standards | discount 10% |

|  |  |
|---|---|
| 40 shelf brackets | discount 10% |
| 25 closet rods | discount 15% |
| 20 aluminum tracks | discount 13% |

_____

**43.** What is the cost of the following hardware that is needed for a small repair job?

| | |
|---|---|
| 60 pounds finish nails | 27¢ per pound with 3% discount |
| 18 sash locks | $2.80 per dozen with 6% discount |
| 84 pounds sash weights | 5¢ per pound with 2% discount |
| 92 sq. ft. screen wire | 7½¢ per sq. ft. with 2% discount |

_____

# Measurement: Direct and Computed

## Unit 18  LINEAR MEASURE

### BASIC PRINCIPLES OF LINEAR MEASURE

*Linear Measure* refers to a measure of length. Usually, this type of measurement is used to measure and order standard types of lumber. For example, studs, molding, rafters, and siding are ordered by length, in linear, lineal, or running feet.

In the English system, the units of linear measure are inches, feet, yards, and miles. In the metric system, the basic unit of linear measure is the metre.

| COMMON ENGLISH LINEAR UNITS | | |
|---|---|---|
| 12 inches | = | 1 foot |
| 3 feet | = | 1 yard |
| 16 1/2 feet | = | 1 rod |
| 5,280 feet | = | 1 mile |

| COMMON METRIC LINEAR UNITS | SYMBOL | VALUE IN METRES |
|---|---|---|
| 1 millimetre | mm | 0.001 |
| 1 centimetre | cm | 0.01 |
| 1 decimetre | dm | 0.1 |
| 1 metre | m | 1.0 |
| 1 dekametre | dam | 10.0 |
| 1 hectometre | hm | 100.0 |
| 1 kilometre | km | 1 000.0 |

(COMPARATIVE SIZES ARE SHOWN)

| 1 METRE |
|---|

| 1 YARD |
|---|

**Metre:** a little longer than a yard (about **1.1** yards or **39.37** inches)
**Millimetre: 0.001** metre    diameter of a paper clip wire
**Centimetre: 0.01** metre    width of a paper clip (about **0.4** inch)
**Kilometre: 1 000** metres    somewhat further than ½ mile (about **0.6** mile)

*Perimeter* is the linear measure of the boundaries of a figure or structure.

The perimeter of a square is the total length of its four sides. Expressed as a formula, $P = 4 \times S$, with $P$ = to perimeter and $S$ = to length of side.

To find the perimeter of a rectangle, use the formula $P = 2 \times (W + L)$; $P$ = perimeter, $W$ = width of rectangle, $L$ = length of rectangle. The use of parentheses in a formula indicates that the mathematical operation within the parentheses must be performed before any other operation.

*Circumference* is the linear measure of the boundary, or arc, of a circle.

To find the perimeter or circumference of a circular object, use the formula $C = \pi D$ or $C = 2\pi r$. $C$ = circumference, $D$ = diameter, $r$ = radius of circle, and $\pi$ (pi) is a Greek letter or symbol used to compute relationships in circular measure. Its value is constant and may be expressed as 3 1/7, 22/7, 3.14, or 3.1416, depending on the degree of accuracy required.

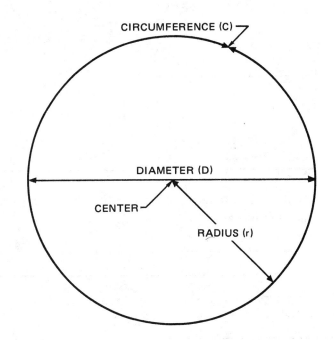

## REVIEW PROBLEMS

1.  Using a rule, determine the length of each line segment to the nearest 1/16 of an inch.

a _____          a _____

b _____             b _____

c _____        c _____

d _____            d _____

e _____         e _____

f _____                             f _____

g _____       g _____

h _____                h _____

i _____               i _____

2.  Using either a caliper or scale, determine the diameter, correct to the
    nearest 1/16 inch, of each bored hole represented.

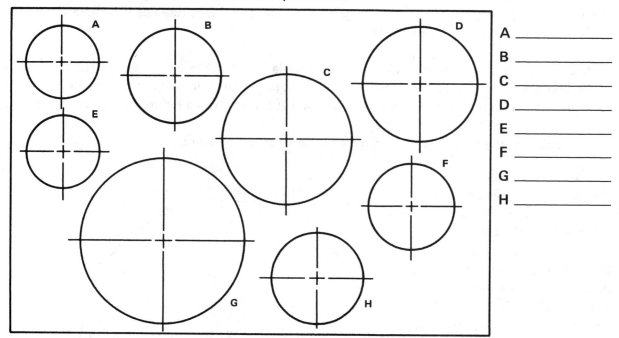

A _____
B _____
C _____
D _____
E _____
F _____
G _____
H _____

3.  Using a metric scale, determine the length of each line segment, correct
    to the nearest millimetre.

a _____   a _____
b _____   b _____
c _____   c _____
d _____   d _____
e _____   e _____

4.  Determine the diameter, correct to the nearest millimetre, of each bored
    hole represented.

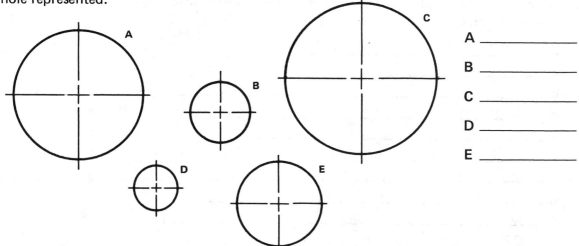

A _____
B _____
C _____
D _____
E _____

**Example of finding the perimeter of a square.**
Find the perimeter of a square with a side of 9 feet.

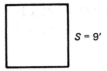

$S = 9'$

1. The formula for the perimeter of a square is $P = 4 \times S.$
2. Substitute 9' for the symbol $S$.     $P = 4 \times 9'$
3. Multiply.     $P = 36'$ Ans.

Find the perimeter of each of the following squares.

5. Side = 18''     _____
6. Side = 34''     _____
7. Side = 25 m     _____
8. Side = 32'     _____
9. Side = 42 km     _____

10. Side = 53'     _____
11. Side = 2'-5''     _____
12. Side = 4'-7''     _____
13. Side = 14'-6''     _____
14. Side = 22'-9''     _____

**Example of finding the perimeter of a rectangle.** Find the perimeter of the illustrated rectangle.

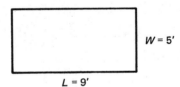

$W = 5'$

$L = 9'$

1. The formula for the perimeter of a rectangle is $P = 2 \times (W + L).$
2. Substitute 5' for $W$ and 9' for $L$.     $P = 2 \times (5' + 9')$
3. Add, then multiply.     $P = 2 \times (14')$
    $P = 28'$ Ans.

Find the perimeter of each of the following rectangles:

15. Width = 11', Length = 14'     _____
16. Width = 12', Length = 16'     _____
17. Width = 150 mm, Length = 220 mm     _____
18. Width = 18'', Length = 32''     _____
19. Width = 21 mm, Length = 22 mm     _____
20. $W$ = 3'-7'', $L$ = 4'-9''     _____
21. $W$ = 5', $L$ = 15'-11''     _____
22. Width = 8 m, Length = 14.5 m     _____
23. Width = 8 m, Length = 14.52 m    
24. $W$ = 18'-9'', $L$ = 38'-6''     _____

**Example of finding the circumference of a circle.** Find the circumference of a circle if the radius is 7'. Use $\pi$ = 3.1416.

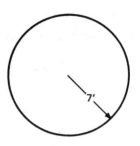

1. The formula for the circumference of a circle is $C = 2 \times \pi \times r$.
2. Substitute 3.1416 for the symbol $\pi$ and 7' for $r$.
3. Multiply.

$$C = 2 \times 3.1416 \times 7'$$
$$C = 6.2832 \times 7'$$
$$C = 43.9824' \text{ or}$$
$$43'\text{-}11\ 25/32''$$

Using $\pi$ = 3.1416, find the circumferences of the following circles:

**Note**:  Solve problems 25–29 correct to the nearest 1/32 of an inch.

25.  Diameter = 32''  _____

26.  Diameter = 4''  _____

27.  Diameter = 13'-0''  _____

28.  Radius = 6'-3''  _____

29.  Diameter = 22'-4''  _____

**Note**:  Solve problems 30–34 correct to the nearest thousandth (3rd decimal place).

30.  Diameter = 16 cm  _____

31.  Diameter = 18 m  _____

32.  Radius = 475 mm  _____

33.  Radius = 14.5 ft.  _____

34.  Diameter = 5 m  _____

Note:  Use this illustration for problems 35–39.

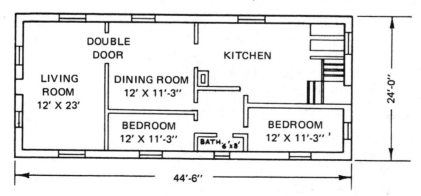

35. What is the perimeter of the living room in the plan shown?  (Do not deduct for openings.)  _____

36. What is the perimeter of the outside walls?  _____

37. How many feet of baseboard are needed for the living room?  (Deduct 3 feet for single door openings, and 5 feet for double door openings.)  _____

38. How many linear feet of shoe base are needed for the two bedrooms? (Deduct 3 feet for each door opening.)  _____

39. How many linear feet each of base and tile cap are used in the bathroom? (Deduct 3 feet for the door opening.)  _____

40. The diameter of a circular plate is nine feet.  What is its circumference?  _____


Find the perimeter of each of the following objects:

41. 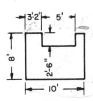  _____  45. 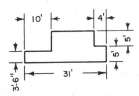  _____

42. 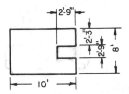  _____  46. 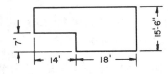  _____

43. 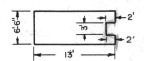  _____  47. 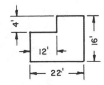  _____

44. 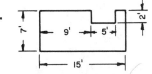  _____  48. 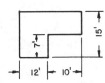  _____

49. What is the perimeter of the semicircular form in the figure? (Express the answer to the nearest thousandth metre.)

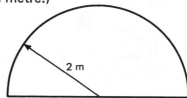

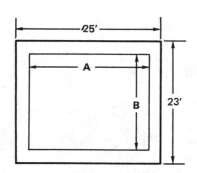

_____

50. A garage footing 18 inches wide, measures 23' x 25' along its outside dimensions. Find the inside dimensions.

A _____

B _____

51. How many linear feet of base and shoe base combined are required for a room 14' x 18' with two doors, allowing 3 feet per door?

_____

52. A blueprint shows outside dimensions of 87' x 87'. How many linear feet of outside wall are there?

_____

53. A new subdivision of a city has outside dimensions of 978' x 978'. What is the distance around it?

_____

54. How many linear feet of 1'' x 4'' will be required to make a form for a circular flower bed if the landscape plan shows a diameter of 22 feet?

_____

# Unit 19   SQUARE MEASURE

## BASIC PRINCIPLES OF SQUARE MEASURE

Area is always measured in units of *square* feet, *square* inches, *square* yards and so forth.  One inch times one inch equals one *square* inch; one foot times one foot equals one *square* foot, and so on.

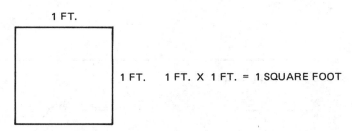

1 FT.

1 FT.    1 FT. X 1 FT. = 1 SQUARE FOOT

To find the area of a rectangle, the width is multiplied by the length.  (All dimensions must be in the same units.  Inches must not be multiplied by feet, or yards by inches, etc.)

**Example:**    Find the area of a rectangle 18″ wide and 3′ long.

First change 18″ to 1 1/2′, then multiply 1 1/2′ by 3′.  The answer will be in *square* feet.

Common area formulas:

Square:    Area = side x side

Rectangle:    Area = length x height (width)

Triangle:  Area = 1/2 x base x altitude (height)

Circle: Area = $\pi$ x radius x radius; or

Area = 0.7854 x diameter x diameter

In the English system, area is measured in square inches, square feet, square yards, and square miles. In the metric system, the basic unit of square measure is the square metre.

| ENGLISH UNITS OF SQUARE MEASURE |
|---|
| 144 square inches  =  1 square foot |
| 9 square feet  =  1 square yard |
| 4,840 square yards  =  1 acre |

| METRIC UNITS OF SQUARE MEASURE |
|---|
| 100 dm$^2$  =  1 m$^2$ |
| 100 cm$^2$  =  1 dm$^2$ |
| 100 mm$^2$  =  1 cm$^2$ |
| 1 000 000 mm$^2$  =  1 m$^2$ |

| SQUARE MEASURE UNITS | SYMBOL | RELATION TO SQUARE METRE |
|---|---|---|
| 1 square metre | $m^2$ | Standard Unit of Area |
| 1 square decimetre | $dm^2$ | 0.01 square metre |
| 1 square centimetre | $cm^2$ | 0.000 1 square metre |
| 1 square millimetre | $mm^2$ | 0.000 001 square metre |

## REVIEW PROBLEMS

Find the area of each of the following squares:

1.  Side = 11'-0''  _____
2.  Side = 14'-0''  _____
3.  Side = 28 m  _____
4.  Side = 42''  _____
5.  Side = 29 cm  _____

6.  Side = 5'-3''  _____
7.  Side = 25'-6''  _____
8.  Side = 38'-3''  _____
9.  Side = 3'-6''  _____
10.  Side = 6'-9''  _____

Find the area of each of the following rectangles:

11.  Width = 5',        Length = 7'  _____
12.  Width = 9',        Length = 11'  _____
13.  Width = 9 m,       Length = 17 m  _____
14.  Width = 19'',      Length = 37''  _____
15.  Width = 32 cm,     Length = 55 cm  _____
16.  Width = 4'-10'',   Length = 8'-2''  _____
17.  Width = 23'-6'',   Length = 38'-6''  _____
18.  Width = 165 cm,    Length = 20 cm  _____
19.  Width = 12'-9'',   Length = 22'-0''  _____
20.  Width = 15'-0'',   Length = 36'-6''  _____

Find the total wall area for each of the following rectangular rooms: *(W =
room width, L = room length, H = wall height. No allowance need be
made for wall openings such as door, windows, etc.)*

21.  *W* = 8',         *L* = 9',         *H* = 7'  _____
22.  *W* = 11',        *L* = 15',        *H* = 10'  _____
23.  *W* = 6 m,        *L* = 8 m,        *H* = 2 m  _____

24. $W = 7'$,      $L = 11'$,      $H = 5'$        _____

25. $W = 5'\text{-}6''$,      $L = 9'\text{-}6''$,      $H = 7'\text{-}9''$        _____

26. $W = 3$ m,      $L = 0.47$ m,      $H = 2$ m        _____

27. $W = 8'\text{-}3''$,      $L = 22'\text{-}6''$,      $H = 6'\text{-}10''$        _____

28. $W = 18'\text{-}6''$,      $L = 25'\text{-}6''$,      $H = 9'\text{-}3''$        _____

29. $W = 6$ m,      $L = 10.5$ m,      $H = 2.9$ m        _____

30. $W = 12'\text{-}9''$      $L = 46'\text{-}9''$,      $H = 7'\text{-}6''$        _____

31. An apartment building contains 14 units, each 18'-6'' x 26'-6''. What is the total floor area?        _____

32. How many square feet of floor area are there in three two-story apartment houses, each of which is 38 feet wide and 76 feet long?        _____

33. A dining room floor measures 12'-0'' wide and 12'-6'' long. At a rate of 6 cents per square foot, what is the cost of sanding the floor?        _____

34. Ten pieces of rigid foam insulation, each measuring 4'-0'' in width and 8'-0'' in length, are purchased at a cost of 35 cents per square foot. What is the total cost of the ten pieces?        _____

Note: Use this diagram
for problems 35 and 36.

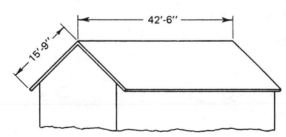

35. Both sides of the roof shown are covered with asphalt shingles. Find the number of square feet of surface covered.        _____

36. How many square feet of surface does this roof contain if the ridge is 33'-9'' long?        _____

37. A bathroom, 6'-6'' wide and 7'-9'' long, has four sidewalls covered with tile to a height of 4'-6''. If an allowance of 15 square feet is made for door and window openings, how many square feet of wall surface are covered with tile?        _____

38. A bathroom, 6'-6" x 7'-8", has a tile floor. Determine the number of square feet covered. (Make no allowance for tub, toilet, etc.)   _____

39. It takes 50 pieces of flashing for a roof. Each piece is 0.5 m wide and 0.8 m long. How many square metres of aluminum flashing are purchased?   _____

40. The four sidewalls of a room 14'-6" wide and 17'-8" long are paneled to a height of 5'-6". A deduction of 27 square feet is made for door and window openings in the room. How many square feet of wall surface are covered?   _____

41. How many square yards of ceiling surface are there in a room 14'-6" x 17'-8"?   _____

42. What is the cost of 32 pieces of paneling 4'-0" wide and 8'-0" long at a unit cost of 20 cents per square foot?   _____

43. Twelve pieces of plywood, each measuring 17 inches wide and 21 1/2 inches long, are used for the backs of cabinets. What is the actual number of square feet of plywood used?   _____

44. Determine the number of square feet of surface in a concrete sidewalk 4'-10" wide and 65'-6" long.   _____

45. How many square feet of cedar are required to line the sidewalls and ceiling of a closet 6 feet wide, 4 1/2 feet deep, and 8 feet high? (Allow 20 sq. ft. for the door.)   _____

46. How many square metres of plywood are required for the tops of 50 tables, each 40 centimetres square?   _____

47. A lawn is planned on the site illustrated. Deduct the area of the house and drive to find the area, in square yards, that is to be seeded.

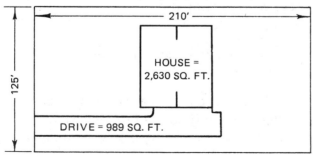

_____

48. At a rate of $1.25 per square foot, what is the cost of the laminated plastic needed to cover the illustrated countertop and back-splash? (End edges are not to be covered.)

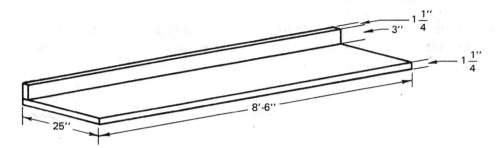

_____

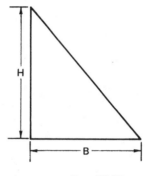 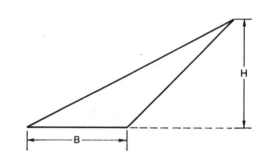 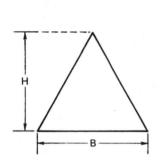

# Unit 20  SURFACE MEASUREMENT— TRIANGLES

## BASIC PRINCIPLES OF SURFACE MEASUREMENT

The procedure for finding the area, or surface measure, of triangles is essentially the same as the procedure for working with rectangles.  The area of a triangle is always one-half the base times the altitude.  ($A = 1/2\ b \times h$, where $b$ = length of base and $h$ = altitude or height)

**Example:**

B = BASE
H = ALTITUDE

## REVIEW PROBLEMS

Find the area of each of the following triangles:

$A$ = Altitude, $B$ = Base

1.  $A = 9'$,     $B = 16'$                    _____

2.  $A = 16''$,    $B = 22''$                   _____

3.  $A = 3'\text{-}10''$, $B = 5'\text{-}5''$             _____

4.  How many square feet of gable are there in the triangular figure?

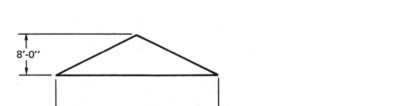

_____

5.  A gable has a span of 50'-0'' and a rise of 10'-0''.  What is the area in square feet?                                            _____

6.  A gable end has a 7'-6'' rise and a span of 30'-6''.  What is its area?    _____

7. How many square feet are covered with 8-inch sheathing if the triangular section of the ramp shown is enclosed on one side only?

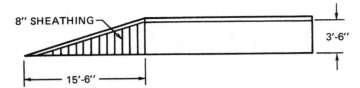

8. What is the area of a pryamid roof shown if the length of one side is 12'-0'' and the common rafter length is 14'-6''? A *pyramid roof* is a square hip roof in the shape of a pyramid.

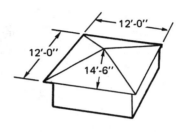

**Note:** Area of a trapezoid = $\dfrac{Base_1 + Base_2}{2}$ X Height

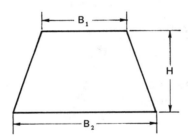

9. The building illustrated is 24 feet wide and 36 feet long. The common rafters are 17 feet long with no tails. The length of the ridge is 12 feet. Find the area of the hip roof. (Hint: Consider the roof as two triangles and two trapezoids, each with altitudes equal to the length of the common rafters.)

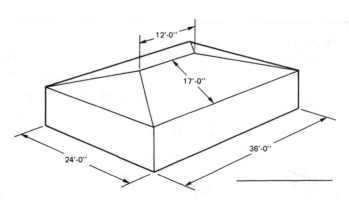

10. A gable roof has a span of 28'-6''. The height of the ridge above the plate is 14'-3''. How many square feet of surface are there in one of the gable ends?

11.  The top of a patio table is in the shape of a triangle which is 40″ on each side. Find the area of the table top, to the nearest square inch. (Hint: First find the altitude of the triangle.)

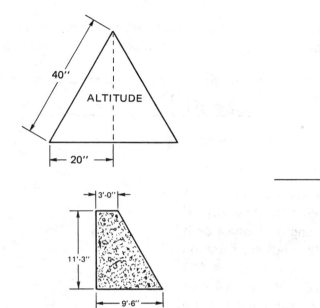

_____

12.  Find the number of square feet of form work required for the end of the concrete retaining wall illustrated.

_____

13.  a. What is the area of the portion marked A of the gambrel roof shown?

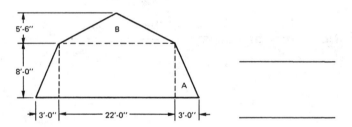

_____

   b. What is the area of portion B in the figure?

_____

   c. How many square feet are covered with siding in the entire gable end shown? Do not make any allowance for the cornice.

_____

Note:  Use this illustration for problems 14-18.

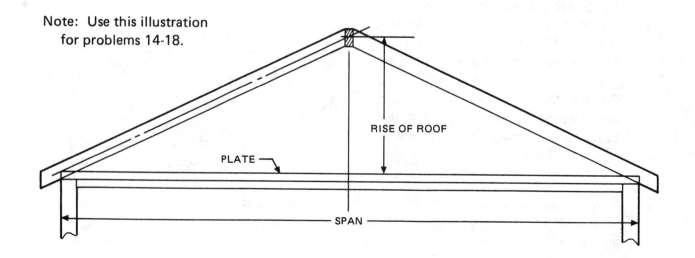

14. What is the area of a gable that has a rise of 6 feet and a span of 18 feet? _____

15. A building is 30 feet wide, and the ridge is 10 feet above the plate line. What is the area of the gable? _____

16. A gable roof has a rise of 4 feet and a span of 16 feet. What is the combined area of the two gable ends? _____

17. How many square feet of wall space are there in the two gable ends of a house if the ridge is 10 feet above the plate line and the building is 36 feet wide? _____

18. A gable roof has a span of 11'-6" and a rise of 7'-6". What is the total area of the ends of the two gables? _____

19. What is the area of each gable end of a house with a span of 33'-0" and a 1/3 pitch? _____

20. a. An unequally pitched roof has a shape similar to the one in the diagram. What is the area of surface A? _____

Note: Use this illustration for problems 20 and 21.

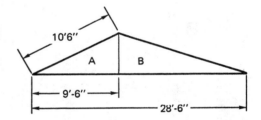

b. What is the area of surface B in the figure. _____

21. a. Find the area of surface A when the given dimensions are doubled.

_____

b. What is the total area of surfaces A and B when the given dimensions are doubled? _____

22. What is the area of the square hip roof shown if the common rafter length is 16'-4"?

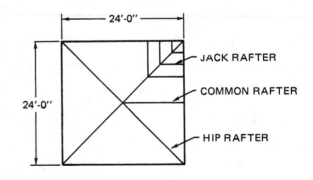

_____

23. A building lot is in the shape of an isosceles triangle with a base of 80'-0" and an altitude of 50'-0". What is its area in square feet? _____

24. a. Find the area of the shaded portions of the hip roof shown if the length of the common rafter is 9'-0''.

   _____

   b. What is the area of the entire roof surface shown?

   _____

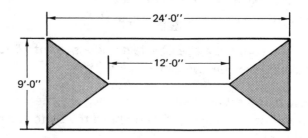

# Unit 21   THE FRAMING SQUARE

## BASIC PRINCIPLES OF THE FRAMING SQUARE

The *framing square,* or *steel square,* is made up of two arms referred to as the blade (or body) and the tongue.

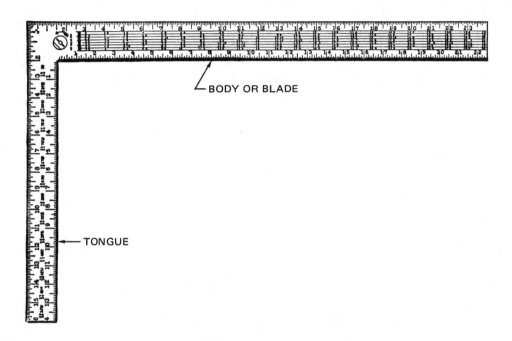

BODY OR BLADE

TONGUE

The *blade* is generally 24 inches long and 2 inches wide. The *tongue* is the smaller arm, normally 16 inches long and 1 1/2 inches wide. The side with the manufacturer's name is the *front* (or *face*), while the other side is the *back*. The carpenter's ability to work with decimals and change them to fractions is very important when using a framing square.

Framing squares contain both scales and tables. Manufacturers of squares show a variety of tables but all have a rafter table.

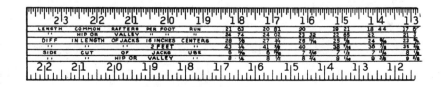

Rafter tables are used in determining the length of Common, Hip, and Valley rafters for roofs of various pitches.  They are used in determining the difference in lengths of Jack rafters and also to aid in making side cuts on Hip, Valley, and Jack rafters.  The problems in this unit are limited to the length of common rafters.

Note:  Use this illustration
for problems 1—6.

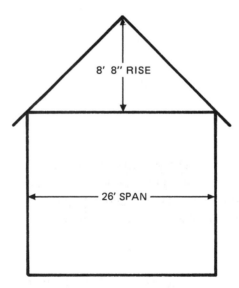

RAFTER TABLE

Study the portion of the framing square shown.  Under the 8-inch mark, the first row of numerals is 14 42.  This indicates that for each foot of run with a rise of 8 inches, the length of a common rafter is 14.42 inches.

**Example:**   Find the length of a common rafter (without allowing for tail or overhang) in the building shown.

The span (26 feet) divided by 2 equals the run (13 feet).  The rise (8 feet 8 inches or 104 inches) divided by the run (13 feet) equals the rise per foot of run (8 inches).  The length of common rafters with a rise of 8 inches per foot of run is 14.42 inches per foot of run.  The total length of this rafter is 13 times 14.42 inches or 187.46 inches.

187.46 inches divided by 12 = 15.6216 feet

15.6216 feet = 15 feet plus 0.6216 x 12 inches

0.6216 x 12 inches = 7.459 inches

7.459 inches = 7 plus 0.459 x $\frac{16}{16}$ inches or 7 $\frac{7}{16}$ inches

Thus, 187.46 inches equals 15 feet 7 $\frac{7}{16}$ inches, the length of the common rafters in the illustration.

## REVIEW PROBLEMS

Using the rafter table, find the length of common rafters in the following problems. Express answers in feet and inches, correct to the nearest 8th of an inch.

1. Run is 7 feet and rise is 10 inches per foot of run. _____

2. Span is 20 feet and rise is 12 inches per foot of run. _____

3. Span is 28 feet and rise is 6 inches per foot of run. _____

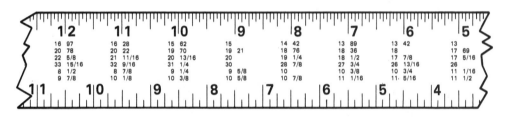

RAFTER TABLE

Note: Use this illustration
for problems 4–6.

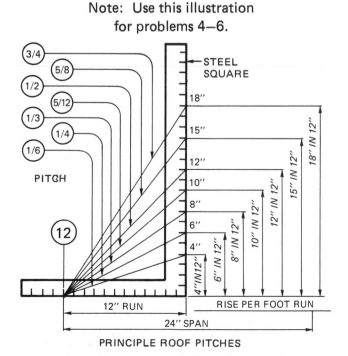

PRINCIPLE ROOF PITCHES

4.  Express 1/3 pitch in terms of rise per foot of run.      _____

5.  Express 1/2 pitch in terms of rise per foot of run.      _____

6.  Find the length of a common rafter with 1/4 pitch and an 8 foot run.      _____

# Unit 22  SURFACE MEASUREMENT—
# IRREGULAR FIGURES

## BASIC PRINCIPLES OF SURFACE MEASUREMENT

There are two fundamental approaches to be used in solving surface measurements of irregular figures. After carefully studying the following methods, select the method best suited to the particular problem you are solving.

Method I

a. Divide the figure into smaller sections.

b. Find the area of each of the smaller sections.

c. Add the areas of the sections to obtain the total area.

Method II

a. Extend the lines which will form a simple geometric figure.

b. Find the area of the large figure.

c. Find the areas of the extra sections.

d. Subtract the areas in step c from the larger area in step b.

## REVIEW PROBLEMS

1. How many square feet of floor surface are there inside the walls of the plan shown? Use method I to find the total area.

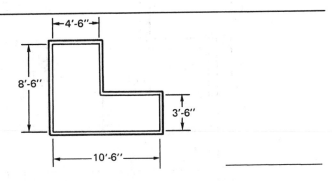

2. Find the area of the plan shown by using method II.

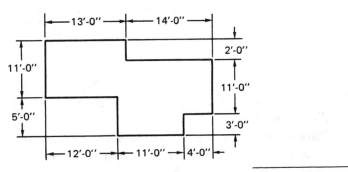

**Note:** Choice of method for the following problems is left to the student.

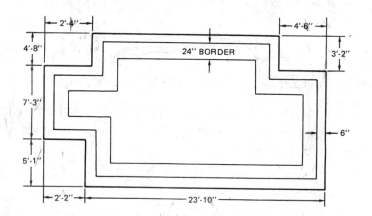

3. The plan shown has walls 6 inches thick. A 24-inch border of hardwood floor is installed around the entire floor.

  a. Determine the area included within the walls.

  b. Find the area of the 24-inch hardwood border.

Note: Use this illustration for problems 4–6.

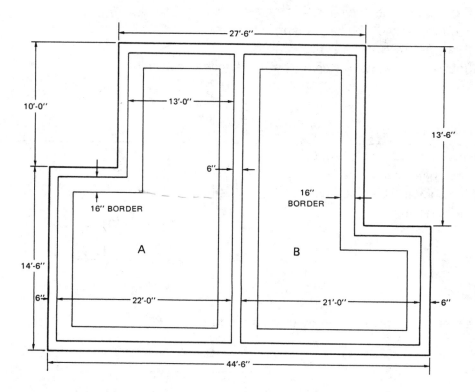

4. How many square feet of floor area are there in room A of the illustration?

5. Determine the floor area to be covered in room B.

6. How many square feet of border surface are there in both rooms if the border is 16 inches wide?

7. How many square inches of surface are there in the end of the form for the retaining wall shown?

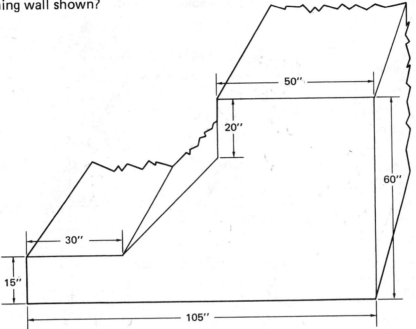

8. A retaining wall, as shown, is 40'-0" long. How many square feet of form work are required for the outside finished face, from top to bottom?

Note: Use this illustration for problems 8–11.

9. What is the area of the end of the retaining wall above the lower grade?

10. Form work is required for the top of the 6-inch steps. How many square feet of form work does it take for the fill side of the retaining wall if it is 40'-0" long?

11. What is the area of the end of the retaining wall below the lower grade line?

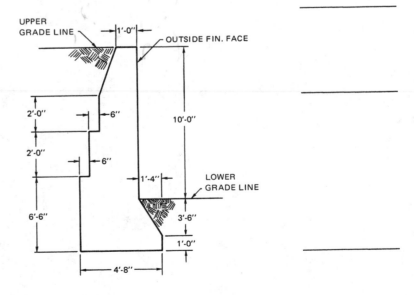

12. In room A of the illustration, how many square yards of lath, correct to the nearest tenth, does it take to lath the ceiling?

13. How many square yards of lath, correct to the nearest tenth, does it take to cover the ceilings of rooms B and C?

14. The walls shown in this figure are 8'-6" high. How many square yards of lath, correct to the nearest tenth, are required for the walls if the openings total 112 square feet?

Note: Use this illustration for problems 12–14.

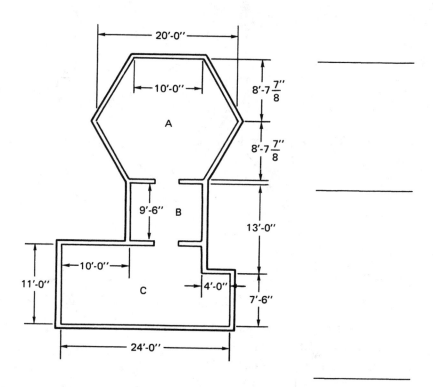

# Unit 23  SURFACE MEASUREMENT— CIRCLES

## BASIC PRINCIPLES OF SURFACE MEASUREMENT

The area of circular figures is found using the formula $A = \pi r^2$, ($r$ is the length of the radius of the circle and $\pi$ is a constant value approximately 22/7 or 3.1416). If remembering this formula is difficult, think of it as $\pi rr$ (pi railroad). If the diameter of the circle is given instead of the radius, remember that the radius is always one-half the diameter ($\frac{D}{2} = r$).

**Example:**   How many square feet of surface area are there in a circular figure with a radius of 8'-6"?

Area = $\pi r^2$
Area = 3.1416 $\times$ 8.5 $\times$ 8.5 = 226.98 sq. ft.

## REVIEW PROBLEMS

Using $\pi$ = 3.1416, find the area of each of the following circles:

Express answers for problems 1–3 correct to the nearest square inch.

Express answers for problems 4–6 correct to the third decimal place.

1.  Diameter = 3'-0"    _____

2.  Diameter = 20"    _____

3.  Radius = 6'-0"    _____

4.  Diameter = 8 m    _____

5.  Diameter = 14 cm    _____

6.  Radius = 7 mm    _____

7.  Cross sections of pipes **A** and **B** are shown. What is the difference, to the nearest square inch, between their inside areas?

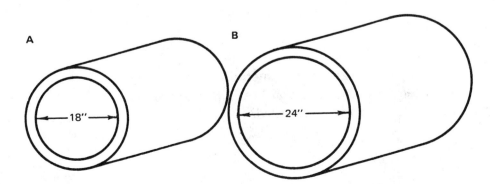

_____

**Note:** Express the answers correct to the nearest thousandth.

8. Determine the floor area in the semicircular bay window shown.

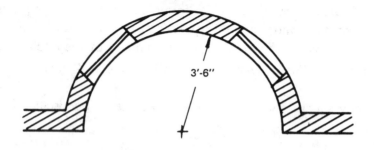

3'-6"

_____

9. How many square yards of surface area are lathed in the ceiling of a semicircular bay window with a radius of 12'-6"?

_____

10. How many square feet of ground area are occupied by the curved runway shown?

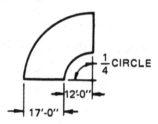

$\frac{1}{4}$CIRCLE

12'-0"

17'-0"

_____

11. How many square yards of lathing are required for a bay window ceiling if the bay is one quarter of a circle in plan and has a radius of 7'-6"?

_____

12. How many square feet of concrete surface are there in the floor of the silo shown?

15'-6"

_____

13. A circular plywood table top with a diameter of 1 metre is covered with laminated plastic. The cover piece is cut from a sheet 1 metre square. How many square centimetres of material are wasted?

_____

14. A stencil is needed to aid in spray painting a playground. If the stencil is cut from a 4' x 8' sheet of plywood, how much waste is created?

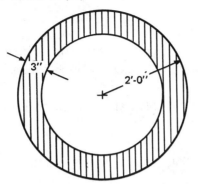

15. A rough floor is laid under the space occupied by the gymnasium track shown. Determine the area occupied by the track.

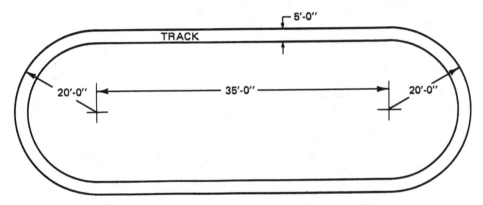

16. Find the cross-sectional area of the wooden block illustrated.

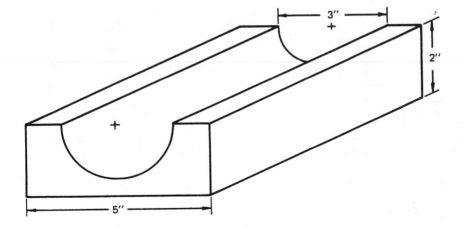

# Unit 24   VOLUME MEASUREMENT— CUBES AND RECTANGULAR SOLIDS

## BASIC PRINCIPLES

Volume is measured in *cubic* inches, *cubic* feet, *cubic* yards, and so forth.  It is never measured in inches, feet, yards, etc.  One inch times one inch times one inch equals one *cubic* inch.

To find the volume of a rectangular solid, first express all dimensions in the same unit of measurement, then multiply width by length by height.

**Example:**  Find the volume of a rectangular solid which is 2 feet wide, 1 yard long and 18 inches high.

Solution:  First change the problem to read 2 feet wide, 3 feet long and 1 1/2 feet high.  The answer will be in cubic feet.

$$V = W \times L \times H$$
$$V = 2' \times 3' \times 1\ 1/2'$$
$$V = 9 \text{ cubic feet}$$

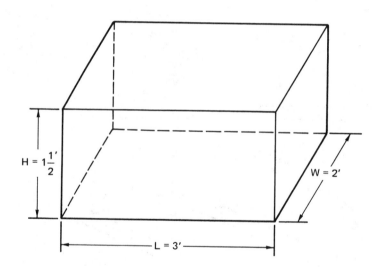

## REVIEW PROBLEMS

Find the volume of the following cubes:

1.  Side = 9'          _____

2.  Side = 14"          _____

3.  Side = 5 cm          _____

4.  Side = 8"          _____

5.  Side = 8'-11"          _____

Find the volume of each of the following rectangular solids:

6.  Length = 6'',      Width = 5'',      Height = 4''      _____

7.  Length = 8 m,     Width = 5 m,    Height = 3 m     _____

8.  Length = 30',     Width = 20',     Height = 12'      _____

9.  Length = 10',     Width = 8',      Height = 2 1/2'   _____

10. Length = 3',      Width = 2',      Height = 6''      _____

11. How many more cubic feet are there in container A than in container B?

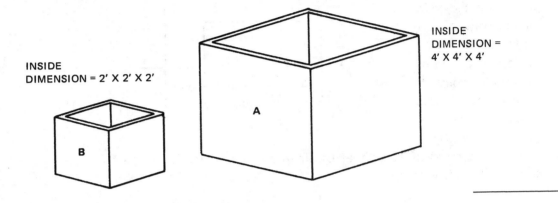

INSIDE DIMENSION = 2' X 2' X 2'

INSIDE DIMENSION = 4' X 4' X 4'

A

B

_____

12. How many cubic metres of space are there in a storage area 8 m wide, 16 m long, and 7 m deep?

_____

13. How many cubic yards of concrete are needed to pour eight footings, 20'' x 20'' x 8''?

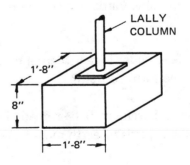

LALLY COLUMN

1'-8''

8''

1'-8''

_____

Note: Use this illustration for problems 14—16.

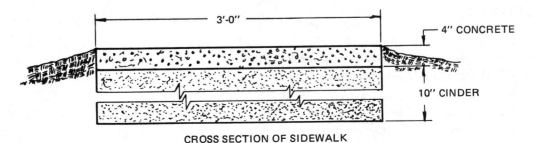

3'-0''

4'' CONCRETE

10'' CINDER

CROSS SECTION OF SIDEWALK

14. The illustration shows a cross section of a concrete sidewalk. How many cubic yards of earth does it displace if the walk is 40'-0'' long?  _____

15. How many cubic yards of concrete are there in the sidewalk shown?  _____

16. How many cubic yards of cinders are required for the sidewalk shown?  _____

17. How many cubic yards of earth must be removed for the excavation shown in the illustration if the depth is 8'-0''? (Note: Excavations are usually computed in terms of cubic yards.)

_____

18. Find the number of cubic feet of air in a room 24 feet long, 16 feet wide, and 7.5 feet high.  _____

19. Find the number of cubic yards of earth to be removed for a basement 45 feet long, 24 feet wide, and 7 feet deep.  _____

20. Determine the number of cubic yards in a concrete column 2 feet square and 10 feet high.  _____

21. A concrete form is 2 feet wide, 7 feet high and 684 feet long. How many cubic yards does the form contain?  _____

22. A large building requires a basement 9 feet deep, 78 feet wide, and 96 feet long. How many cubic yards of earth must be removed?  _____

23. How many cubic feet of concrete are required for a retaining wall 40 feet long, 6 feet high, and 1 1/2 feet thick?  _____

24. How many cubic feet are taken up by a stairwell 9'-6'' long, 3'-4'' wide, and 8'-6'' high?  _____

25. How many cubic feet of storage space are there in a closet 6 feet long, 2 1/2 feet deep, and 7 feet 6 inches high?  _____

26. A storage space in a basement is 8'-6'' x 6'-8'' x 8'-8''. What is the number of cubic feet of storage space?  _____

27. The illustration shows a section of a concrete foundation wall and footing. How many cubic yards of concrete does it contain?

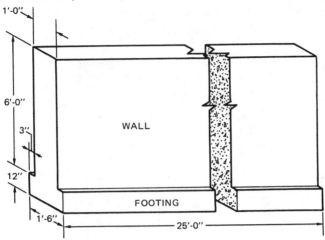

28. An apartment house requires a concrete foundation wall **9 inches** thick and 3 feet high. The size of the building is **38' x 82'**.

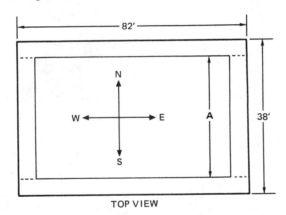

TOP VIEW

    a. How many cubic feet of concrete are needed for both the north and south walls?

    b. Find missing dimension **A**.

    c. How many cubic feet of concrete are needed for the east and west walls?

    d. How many cubic feet of concrete are there in the entire foundation?

29. A building is **28 feet** wide and **44 feet** long. The excavation for the footing is **18 inches** wide and **6 inches** deep. How many cubic feet of earth must be removed in digging a trench for just the footing? Assume the excavation for the foundation to have been previously dug.

30. A foundation wall, 25 feet by 44 feet 8 inches, is built of rubble stone. It is 8 feet deep and 16 inches thick. If no footing is included, how many cubic yards, to the nearest tenth, of stone are required for the job?   _____

31. The foundation walls of a house have outside dimensions of 25'-0" x 44'-6". The walls are 6'-0" high and 8" thick. The footings require 270 cubic feet of concrete. Determine the total number of cubic yards of concrete required for the foundation walls and footings.   _____

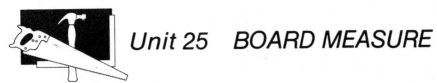

# Unit 25  BOARD MEASURE

## BASIC PRINCIPLES OF BOARD MEASURE

A *board foot* (bd. ft.) is defined as the equivalent of a piece of wood measuring one foot (1'-0'') wide, one foot (1'-0'') long and one inch thick. In each of the sketches, the number of board feet is shown.

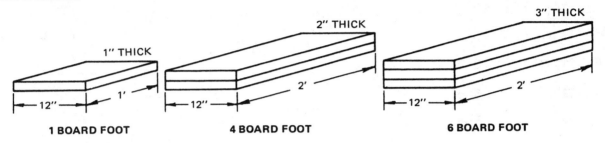

| 1 BOARD FOOT | 4 BOARD FOOT | 6 BOARD FOOT |

To calculate the board measure in any quantity or piece of lumber use the formula:

$$\text{Board feet} = T'' \times W' \times L'$$

in which $T$ = thickness (expressed in inches), $W$ = width (expressed in feet) and $L$ = length (expressed in feet).

**Example:**    To find the number of board feet in the piece of lumber shown,

Board feet   $= T \times W \times L$

Board feet   $= \dfrac{1 \times 6 \times 4}{12} = 2$ bd. ft.

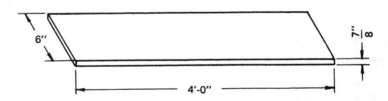

In the board feet problems in this text, use the rough stock dimensions. *Rough stock* is lumber that is not planed or surfaced. *Dressed stock* is lumber that has been surfaced. In dressing a board the width and thickness can be reduced by 1/2 inch. For example, a 2 x 4 will measure about 1 1/2 inches by 3 1/2 inches when dressed.

**Note:** Board measure problems are figured using standard dimensions for thickness and width. Lumber less then 1'' thick is counted as a full inch. Standard *thicknesses* are 1'', 5/4'', 6/4'', 2'', 3'', 4'', 6'', etc. Standard *widths* are 2'', 3'', 4'', 5'', 6'', 8'', 10'', and 12''. For odd widths or thicknesses, count the next standard dimension above the exact dimension desired.

## REVIEW PROBLEMS

Find the number of board feet in each of the following quantities:

1.   5 pcs. of 1'' x 6'' x 18'                 _____

2.   24 pcs. of 1'' x 8'' x 14'               _____

3.   6 pcs. of 2'' x 3'' x 12'                _____

4.   14 pcs. of 1'' x 4'' x 18'              _____

5.   34 pcs. of 2'' x 4'' x 16'              _____

6.   54 pcs. of 1'' x 3'' x 20'              _____

7.   28 pcs. of 2'' x 6'' x 10'              _____

8.   32 pcs. of 1'' x 10'' x 16'             _____

9.   22 pcs. of 2'' x 12'' x 22'             _____

10.  62 pcs. of 1'' x 10'' x 18'            _____

Find the number of board feet in the following quantities:

11.  18 pieces of 1/2'' x 4'' x 16'           _____

12.  62 pieces of 3/4'' x 9 1/4'' x 14'       _____

13.  84 pieces of 5/8'' x 10'' x 16'          _____

14.  77 pieces of 3/4'' x 3 1/2'' x 18'       _____

15.  14 boards, 1'' thick, 12'' wide, 10' long       _____

16.  8 boards, 1 1/2'' thick, 22'' wide, 16' long     _____

17.  6 planks, 2'' thick, 10'' wide, 14' long        _____

18.  10 pieces, 2'' x 6'' x 12', redwood sill         _____

19.  15 pieces, 1'' x 8'' x 14', subfloor stock       _____

20.  25 pieces, 3/4'' x 3'' x 12', Douglas fir        _____

21.  1 piece, 2'' x 3'' x 20', Douglas fir            _____

22.  60 linear feet of 1'' x 4'' redwood             _____

23.  250 linear feet of 1'' x 6'' sugar pine          _____

24.  150 linear feet of 2'' x 3'' redwood            _____

25.  72 linear feet of 1'' x 3'' Douglas fir          _____

26. Determine the total number of board feet of select sugar pine in the following list:

    10 pieces of 5/4'' x 4'' x 16'-0''
    6 pieces of 5/4'' x 10'' x 18'-0''
    1 piece of 5/4'' x 12'' x 12'-0''

    _____

Note: Use this illustration
for problems 27—35

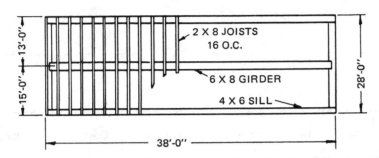

27. The building shown requires two 4 inch by 6 inch sills, each 38 feet long. They are constructed from four pieces of 4 inch by 6 inch timber, each 20 feet long. This allows for a splice 2 feet long in the center of each sill. How many board feet of lumber must be ordered?

    _____

28. Sill stock is sold in multiples of 2'-0'' in length. Find the number of board feet of lumber to be ordered for the two sills that run at right angles to the floor joist if 1 foot in length is added for splicing.

    _____

29. Determine the board feet of stock to be ordered for each floor joist running across the 13'-0'' span. (Floor joists are sold in multiples of 2'-0'' in length.)

    _____

30. How many board feet of stock are to be ordered for all of the floor joists for the 13'-0'' span? (To find the number of joists required, divide the distance by the spacing, 16'' o.c., and add 1 for a starter.)

    _____

31. Determine the board feet of stock to be ordered for the two 4'' x 6'' sills that run parallel to the floor joist. (Make no allowance for splicing.)

    _____

32. Find the number of board feet that must be ordered for each floor joist running across the 15'-0'' span.

    _____

33. How many board feet of lumber are ordered for all of the floor joists running across the 15'-0'' span?

    _____

34. Find the number of board feet of subflooring to be ordered for the entire floor surface if 1'' x 6'' matched boards are to be used. (The number of board feet to be ordered is found by determining the area and adding 20% for matching and waste.)

    _____

35. How many board feet of 1'' x 3'' oak flooring must be ordered for the entire floor, as shown, if an allowance of 38% is made for waste and matching?    _____

36. Complete the following bill of material, and determine the total cost of lumber needed to construct a small table. (Round off the cost of each piece of lumber to the nearest cent.)

### BILL OF MATERIAL

Project _____Table_____          Name _____

| Kind of Wood | # Pcs. | Th. | Wdth. | Lgth. | Bd. Ft. | Cost/Bd. Ft. | Cost | Description |
|---|---|---|---|---|---|---|---|---|
| Maple | 1 | 1'' | 18'' | 30'' | | $0.96 | | Top |
| Maple | 2 | 1'' | 2 ½'' | 12'' | | $0.96 | | End Rail |
| Maple | 2 | 1'' | 2 ½'' | 22'' | | $0.96 | | Side Rail |
| Maple | 4 | 2'' | 2'' | 17'' | | $1.10 | | Legs |
| Maple | 2 | 1/2'' | 1 ½'' | 13'' | | $1.10 | | Bottom, Stretcher, End |
| Maple | 2 | 1/2'' | 1 ½'' | 23'' | | $1.10 | | Bottom,Stretcher,Side |

Total   Cost   _____

# Unit 26  VOLUME MEASUREMENT— CYLINDERS

## BASIC PRINCIPLES OF VOLUME MEASUREMENT

The procedure for finding the volume of a cylinder calls for the use of another new formula: $V = \pi r^2 h$, where $r$ = the radius of the circular end, $\pi$ = 3.1416, and $h$ = the height of the cylinder when viewed with the circular surface as the base.  All measurements must be in the same units before proceeding to multiply.

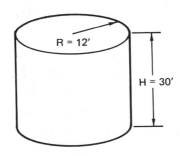

**Example:**   Find the volume of a circular cylinder with a radius of 12 feet and height of 30 feet.

$V = \pi r^2 h$
$V = 3.1416 \times 12' \times 12' \times 30'$
$V = 3.1416 \times 144$ sq. ft. $\times 30$ ft.
$V = 3.1416 \times 420$ cu. ft.
$V = 1434.72$ cu. ft.

## REVIEW PROBLEMS

**Note:**  Let $\pi$ = 3.1416.  Express all answers correct to the nearest hundredth.

1. A cross section of a cylindrical cistern is shown.  What is the volume of the cistern?

2. How many cubic yards of earth are excavated for a concrete cistern that has a diameter of 9'-6'' and a depth of 10'-0'' as outside dimensions?  An allowance of 18 inches is made around the walls for form work.  The top of the cistern is 4'-0'' below the grade.

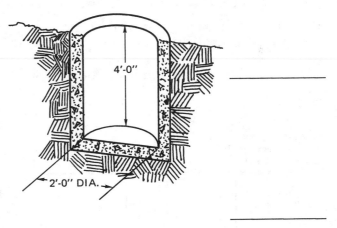

3. What is the capacity, in gallons, of a wooden storage tank that has an inside diameter of 16'-0'' and an inside height of 18'-9''?  One cubic foot contains 7.481 gallons.

Note:  Use this illustration for problems 4 and 5.

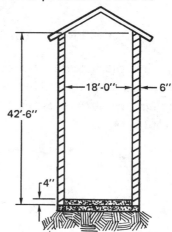

4.  What is the capacity, in cubic feet, of the circular silo shown?

_____

5.  How many cubic yards of concrete are needed for the wall of this silo?

_____

6.  A contractor must estimate the excavation and rock fill needed for two circular dry wells.  One well has a diameter of 5'-0'' and a depth of 6'-6''; the other has a diameter of 5'-6'' and a depth of 7'-0''.  How many cubic yards of rock fill are needed?

_____

7.  A concrete silo base has a diameter of 17'-6'' and a depth of 8''.  How many cubic yards of concrete are needed for the silo base?

_____

8.  A loading dock is supported by two solid concrete piers.  The piers are each 2'-0'' in diameter and 10'-0'' high.  How many cubic feet of concrete do the two piers contain?

_____

9.  Sections of wrought iron pipe, having an inside diameter of 4'', are filled with concrete and used for posts to support the main floor girder in the basement of a residence.  If these posts are 8'-0'' long, and there are 6 in all, how many cubic feet of concrete are required for the job?

_____

10.  How many cubic yards of concrete are needed for a silo foundation that is 14'-6'' in diameter and 2'-6'' deep?

_____

Note:  Use this illustration for problems 11–15.

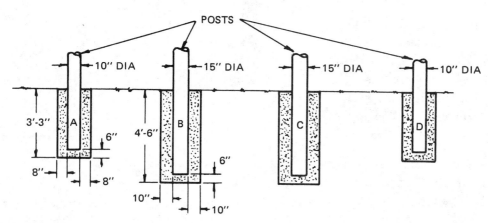

11. The posts in the illustration are set in concrete. All posts and concrete casings are cylindrical. How many cubic feet of earth are excavated for setting both identical posts A and D?  _____

12. How many cubic feet of concrete are needed to pour around both posts A and D?  _____

13. The two posts B and C are alike. How many cubic feet of earth are removed for setting both posts B and C?  _____

14. How many cubic feet of concrete are needed to pour around post B?  _____

15. Determine the number of cubic feet of concrete needed to set the four posts shown if posts A and D are changed to 8 inches in diameter, and posts B and C are changed to 12 inches in diameter. All other dimensions remain unchanged.  _____

16. A contractor has to estimate the rock fill for four circular dry wells. If two of these wells are 4'-0" in diameter and 4'-6" deep, and the remaining two are 3'-6" in diameter and 5'-0" deep, how many cubic yards of fill are required for the wells?  _____

17. Determine the number of square feet of wall surface on the inside of a wooden silo that has a capacity of 3,392.7 cubic feet and a height of 30 feet.  _____

18. What is the volume, in cubic inches, of the pattern block shown?

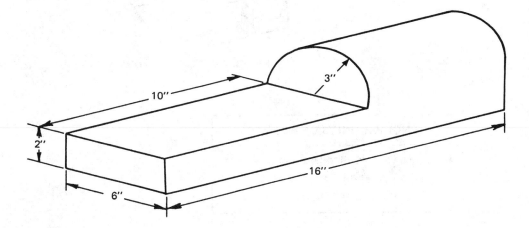

_____

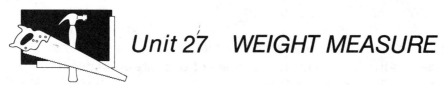

# Unit 27   WEIGHT MEASURE

## BASIC PRINCIPLES OF WEIGHT MEASURE

It is important for carpenters to solve problems involving weight. Loads must be kept within safe bearing capacities of various soils. When hauling materials, the safe capacities of the truck must not be exceeded. When designing a structure, the live load and dead load must be figured to use beams of the most efficient size. This unit contains problems carpenters will encounter in their daily work, including changing from one unit of measurement to another.

In the English system, common units of weight and their equivalents are ounces, pounds, and tons.

The basic metric unit of weight is a gram.

Gram:  about the weight of a paper clip.

Kilogram:  1 000 grams; a little more than 2 pounds (about 2.2 pounds)

1 pound is slightly less than 1/2 kilogram

1 pound = 0.45 kilogram.

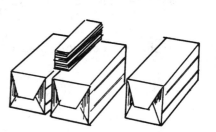

1 KILOGRAM          1 POUND

| ENGLISH UNITS OF WEIGHT |
| --- |
| 1 ounce (oz.)  =  Standard unit of weight |
| 1 pound (lb.)  =  16 ounces |
| 1 ton (T.)  =  2,000 pounds |

| METRIC UNITS OF WEIGHT | | |
| --- | --- | --- |
| Weight Measure | Symbol | Relation to Gram |
| gram | g | Standard unit of weight |
| kilogram | kg | 1 000 grams |
| metric ton | t | 1 000 000 grams |

## REVIEW PROBLEMS

Express the following quantities as equivalent weights in the units indicated:

1. 80 pounds, 13 ounces as ounces                                    _____

2. 350 ounces as pounds                                              _____

3. 155 tons, 300 pounds as pounds                                    _____

4. 2.5 kilograms as grams                                            _____

5. 300 grams as kilograms                                            _____

6. If asphalt shingles weigh 235 pounds per square (100 square feet), how much does the quantity of shingle, needed to cover the area shown, weigh?

_____

7. The safe bearing capacity of a soil is 2,500 pounds per square foot. What area of footing is required to sustain a load of 60,000 pounds?          _____

8. The total weight of a tank when completely filled is 35,000 pounds. How much per square foot is carried by a footing 22 square feet in area?          _____

9. If a live load of 70 pounds per square foot is added to the dead load, what is the live load on a floor measuring 40' x 20'?          _____

10. The type of concrete generally used weighs about 145 pounds per cubic foot. Estimate the weight of a concrete wall 20 feet long, 6 feet high, and 14 inches thick.          _____

11. I beams are ordered and sold by the size and weight per linear foot. At 40 pounds per foot, how much does a 10'' x 4 3/4'' I beam, 24 feet long, weigh?          _____

12. How much will 24 feet of a 10'' x 4 3/4'' I beam weigh if it weighs 35 pounds per linear foot?          _____

13. The weight of a certain quality of sheet steel is 487.7 pounds per cubic foot.  Find the weight of a plate of this steel which is 1/8 inch thick and has 24 square feet of surface on each face.  _____

14. Water weighs 1 000 kilograms per cubic metre. Find the weight of the water contained in the holding tank shown if the tank is filled to the level indicated.

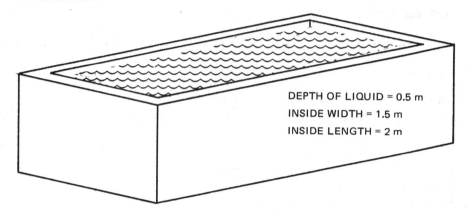

DEPTH OF LIQUID = 0.5 m
INSIDE WIDTH = 1.5 m
INSIDE LENGTH = 2 m

_____

15. Sheet copper is ordered by size and weight per foot but is sold only by the actual weight of the material.  A carpenter orders a sheet of copper weighing 16 ounces per square foot.  The sheet was 16'' x 16'-0''. How many pounds does the sheet weigh?  _____

16. Find the weight, in pounds, of a sheet of 14-ounce copper that is 18 inches wide and 24 feet in length.  _____

17. A contractor purchased a half ton of 8d common nails for $310.  What was the unit price per box (50 pounds)?  _____

18. It requires 10 pounds of 3d nails for each 100 square yards of lath. How many boxes (50 pounds) of nails are required for 2,000 square yards of lath?  _____

19. In house framing, 20 pounds of 8d common wire nails are estimated for each 1,000 board feet of sheathing or subflooring.  How many pounds of nails should be ordered for 15,000 board feet of sheathing?  _____

20. Sand is estimated by the cubic yard but in certain localities it is sold by the ton.  Find the weight of 5 cubic yards of washed sand, allowing 95 pounds per cubic foot of material.  _____

21. When wood shingles are laid 4 1/2 inches to the weather, an allowance of 5 3/4 pounds of 4d nails is made for each square of shingles. How many pounds of shingle nails must be ordered to lay 22 1/2 squares of shingles at this exposure?  _____

22. At $7.25 per ton, find the cost of 15 cubic yards of washed and screened gravel, weighing 106 pounds per cubic foot.  _____

23. How many tons of washed gravel, weighing 105 pounds per cubic foot, will a truck carry if the inside body measurements are 5 feet 6 inches wide, 14 feet long, and 2 feet 6 inches high? (The material is leveled off to the top of the body.)

_____

24. Determine the total weight of the poured concrete wall and footing illustrated if the concrete used weighs 145 pounds per cubic foot.

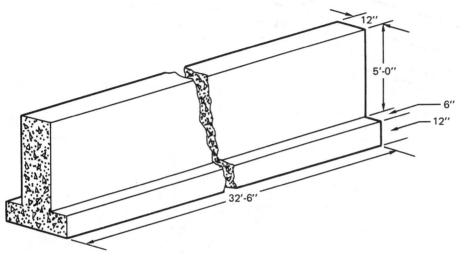

_____

25. If roll roofing weighs 90 pounds per roll and covers an area of one square (10' x 10' or its equivalent), find the weight of roll roofing used to cover both sides of a gable roof, each side measuring 42 feet long and 16 feet wide.

_____

26. A shed roof, 16' x 28', is covered with asphalt shingles weighing 235 pounds per square, with 4 1/2 pounds of nails used per square. Find the total weight of shingles and nails that are applied to the roof.

_____

27. Find the weight, correct to the nearest hundredth, of the steel angle bracket shown if the material weighs 0.28 pounds per cubic inch.

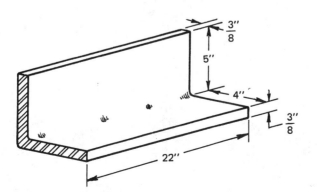

_____

28. What is the total weight of 15 pieces of bar stock if each piece has dimensions as indicated?  (Note:  1 square inch of stock weighs 30 pounds per linear foot.)

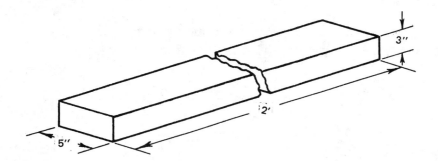

# Ratio and Proportion

## SECTION 6

## Unit 28 RATIO

### BASIC PRINCIPLES OF RATIO

A ratio is a comparison of two quantities, commonly stated as a fraction. Like all fractions, ratios may be reduced to lowest terms by dividing both the numerator and denominator by the same number. The numerator and denominator may also be multiplied by the same number.

**Example:** $\dfrac{25}{35} = \dfrac{25 \div 5}{35 \div 5} = \dfrac{5}{7}$

$\dfrac{1/3}{5} = \dfrac{1/3 \times 3}{5 \times 3} = \dfrac{1}{15}$

**Example:** In five hours a carpenter and his helper can install 100 linear feet of metal eaves, gutters and downspouts. This would be a ratio of 5/100, expressed in simplest terms as 1/20.

$\dfrac{5}{100} = \dfrac{5 \div 5}{100 \div 5} = \dfrac{1}{20}$

### DEFINITIONS

The terms pitch, rise, run, and span are used in connection with roof layout and construction. These terms are shown in their correct relationship in the illustration.

*Pitch* is the ratio of the rise to the span. Expressed as a formula,

$$\text{Pitch} = \frac{\text{Rise}}{\text{Span}}$$

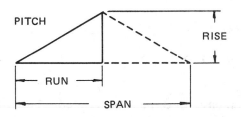

If, for example, the outside points of the plates of a building are 24 feet apart, and the ridge is 8 feet above the plate line, the pitch is equal to 8/24 or 1/3. This is expressed as 1/3 pitch or 1:3.

In the shed roof shown, the *span* is twice the run. The pitch of this type of roof is the ratio of the rise to twice the run, or

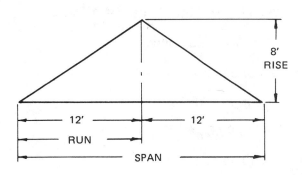

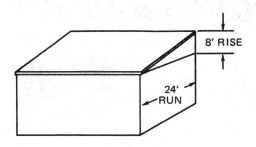

$$\text{Pitch} = \frac{\text{Rise}}{2 \times \text{Run}} \quad \text{or} \quad \text{Pitch} = \frac{\text{Rise}}{\text{Span}}$$

For the roof shown, the pitch equals

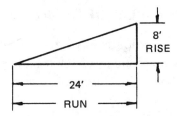

$$\frac{8'}{2 \times 24'} = \frac{8}{48} = \frac{1}{6}$$

Expressed in its simplest form:

$$P = 1/6 \text{ or } 1:6$$

## REVIEW PROBLEMS

Express each of the following ratios in simplest form:

1.  15:25    _____

2.  90:360    _____

3.  7:42    _____

4.  18:36    _____

5.  60:25    _____

6.  36:24    _____

7.  $2: \dfrac{1}{4}$    _____

8.  3:0.7    _____

9.  0.01:0.4    _____

10. $\dfrac{1}{6} : \dfrac{5}{12}$    _____

11. 0.01:0.1    _____

12. 150:25    _____

In each case, find the ratio of the first quantity to the second:

13.  1 yard to 2 feet    _____

14.  8 hours to 1 day    _____

15.  750 cm$^3$ to 1 litre (1 litre = 1 000 cm$^3$)    _____

16.  12 ounces to 5 pounds    _____

17.  9 months to 1 year    _____

18.  600 m to 1 km    _____

19.  35¢ to $2.50    _____

20.  1,200 pounds to 1 ton    _____

21. In the illustration, the rise is 8'-0" and the run of each rafter is 12'-0". What is the pitch.

Note: Use this illustration for problems 21 and 22.

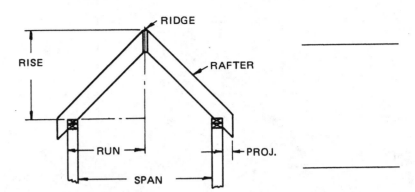

_____

22. If, on the same type of roof, the rise is 12'-0" and the run of each rafter 12'-0", what is the pitch?

_____

Note: Use this illustration for problems 23 and 24.

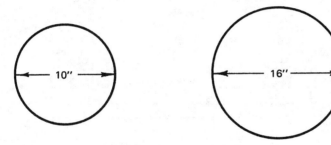

23. What is the ratio of the diameters of the two illustrated circles if the dimensions are as indicated?

_____

24. What is the ratio of the areas of the two circles?

_____

25. On the shed roof shown, the run is 12'-0" and the rise is 3'-0". What is the pitch?

Note: Use this illustration for problems 25–27.

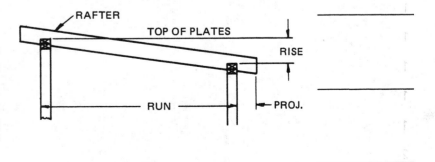

_____

26. What is the pitch of a shed roof having a rise of 2 m and a run of 8 m?

_____

27. What is the pitch of a shed roof having a run of 8'-0" and a rise of 2'-0"?

_____

28. What is the ratio of the lengths of the rectangles shown?

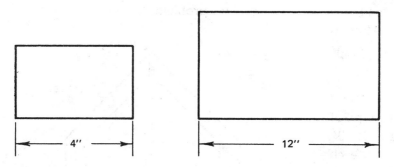

Note: Use this illustration
for problems 29—32.

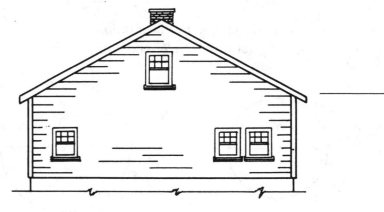

29. The building shown has a gable roof. What is the roof pitch if the rise is 8'-0'' and the span is 24'-0''?

30. What is the pitch of a gable roof if the span is 28'-0'' and the rise is 14'-0''?

31. What is the roof pitch if the span is 6 m and the rise is 2.5 m?

32. If the run of the gable roof is 13'-6'' and the rise is 9'-0'', what is the roof pitch?

33. The gable roof of a small building has a run of 24'-0'' and a rise of 8'-0''. What is the pitch?

34. The figure shows a garage which has a hip roof. What is the roof pitch if the run is 12'-0'' and the rise is 8'-0''?

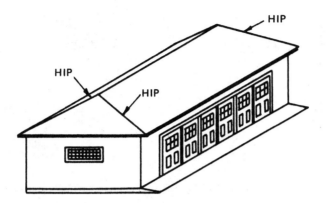

_____

35. What is the roof pitch for a hip roof when the span is 24'-0'' and the rise is 6'-0''?

_____

# Unit 29   PROPORTION

## BASIC PRINCIPLES OF PROPORTION

A proportion is a statement of two equal ratios such as 1/2 = 4/8. In many practical situations, three numbers of a proportion are known but the fourth is unknown.

The four terms of a proportion are named by their location:

$$\text{1st term} \longrightarrow \frac{1}{2} = \frac{4}{8} \longleftarrow \text{3rd term}$$
$$\text{2nd term} \qquad\qquad \longleftarrow \text{4th term}$$

Some interesting characteristics of proportions should be understood before solving problems in this unit.

1. Product of the 1st and 4th terms equals the product of the 2nd and 3rd terms.
2. The 1st and 4th terms may be interchanged.
3. The 2nd and 3rd terms may be interchanged.
4. Both ratios may be inverted, but not just one.

A ratio is a comparison of two quantities by division.

**Example:**    $3/6$, $\frac{3}{6}$, $6\overline{)3}$ , 3:6 all represent the same concept.

## REVIEW PROBLEMS

In each of the following, find the missing quantity.

1. 1:2 = 8:? _____

2. 4:3 = ?:22.5 _____

3. 5:? = 45 yd.:63 yd. _____

4. 12.25:50 = 61.25:? _____

5. $\dfrac{72}{48} = \dfrac{660 \text{ cu. yd.}}{?}$ _____

6. ?:54 = 28:42 _____

7. 33:50 = 11:? _____

8. 75:? = 25:40 _____

9. $16\frac{1}{2} : 24\frac{3}{4} = 40:?$ _____

10. $33\frac{1}{3} : 50 = ?:200$ _____

11. $16:$64 = X:$4$ _____

12. X:85 = 10:17 _____

13. 24:X = 15:40 _____

14. X:75 yd. = $15:$5 _____

15. 1:5 = 1.2 cm:X _____

16. 8.12:X = 112:41 _____

17. An earth embankment rises 1 1/2 feet on every foot of level ground. How much will the embankment rise for 18 feet of level ground? _____

18. If 75 pounds of nails cost $30, what do 125 pounds cost at the same rate? _____

19. If one person does a piece of work in 4 days which a second person can do in 7 days, how long will it take the first person to do a job the second can do in 63 days?

_____

20. The lengths of the two rectangles shown below are proportional to their widths. What is the length of the smaller rectangle?

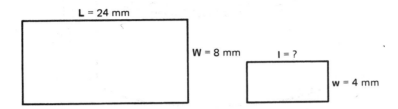

_____

21. How many pounds of nails are required for 1,852 square feet of metal lath if 8 pounds are used for each 1,000 square feet?

_____

22. For certain plaster work 1 1/3 cubic yards of sand are needed for each 100 square yards. How much sand is needed for 4,275 square feet?

_____

23. Determine the quantity of priming paint needed for 3,500 square feet if one gallon covers 750 square feet.

_____

24. If 6 square feet of 8-inch brick wall with 3/8-inch joists contains 87 bricks, find the number of bricks needed for 130 square feet.

_____

25. White pine weighs 25 pounds per cubic foot; steel, 490 pounds per cubic foot. Find the ratio of their weights.

_____

26. Find the number of pounds of nails required for 3,570 square feet of metal lath if each thousand square feet of lath requires 8 pounds of nails.

_____

27. A 24-inch pulley running at 180 revolutions per minute drives a 16-inch pulley. How many revolutions per minute does the smaller pulley make?

_____

28. A 14-centimetre pulley makes 240 revolutions per minute and drives a larger pulley making 210 revolutions per minute. What is the diameter of the larger pulley?

_____

29. Determine the diameter of the small pulley in the illustration.

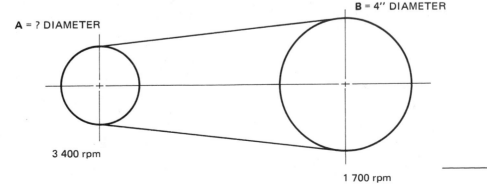

_____

30. The reservoir pictured contains 7 580 litres of water when completely filled. Determine the number of litres it contains when the depth of the water is 2.1 m.

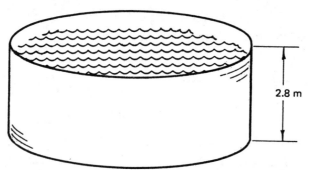

# Powers and Roots

## Unit 30   APPLICATION OF EXPONENTS IN FORMULAS

### BASIC PRINCIPLES OF POWERS

An exponent is a small figure written to the right and slightly above a given quantity. Exponents are used as a convenient way to indicate the number of times a quantity is to be multiplied by itself.

**Example:**   $5^2 = 5 \times 5 = 25$
$10^3 = 10 \times 10 \times 10 = 1{,}000$

Formulas used in this unit involve the use of exponents.

**Examples:**   Area of a square    $A = s^2$
Volume of a cube   $V = e^3$
Area of a circle    $A = \pi r^2$

### REVIEW PROBLEMS

1.  Find the area of the 8-inch square shown.
    $(A = s^2)$

    8″ ◻ 8″

    _____

2.  Find the area of a square whose side is 4′-3″ long.

    _____

Find the area of each of the following squares:

3.  Side = 2″          _____

4.  Side = 5 cm        _____

5.  Side = 6″          _____

6.  Side = 10″         _____

7.  Side = 12″         _____

8.  Side = 1 km        _____

9.  Side = 3′-0″       _____

10.  Side = 7 mm       _____

11.  Side = 2.5″       _____

12. Find the volume of the 8-cm cube shown. $(V = s^3$ or $V = s \times s \times s)$

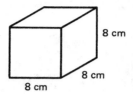

8 cm

8 cm

8 cm

_____

13. Find the volume of a cube whose side is 3'-9'' long.

_____

14. A circle has a diameter of 8 cm. Use $\pi = 3.1416$ to determine the area. $(A = \pi r^2)$

_____

15. Find the area of a circle with a radius of 2'-7''.

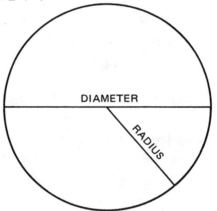

DIAMETER

RADIUS

_____

Find the area of each of the following circles:

16. Radius = 7 cm     _____

17. Diameter = 2 m     _____

18. Radius = 2'-0''     _____

19. Diameter = 9.6 inches     _____

20. Find the volume of a cylinder with a radius of 8'' and a height of 14''. $(V = \pi \times r^2 \times h)$

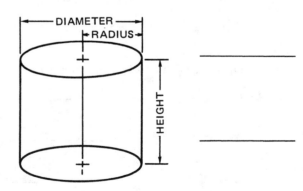

DIAMETER

RADIUS

HEIGHT

_____

21. Find the volume of a cylinder with a diameter of 16'' and a height of 2'-0''.

_____

Find the volume of each of the following cylinders:

22. Radius = 7'',          Height = 4''

_____

23. Radius = 14'-0'',          Height = 7''

_____

24. Diameter = 6 cm,      Height = 10 cm      _____

25. Diameter = 7″,        Height = 12′-0″     _____

26. Diameter = 1 m,       Height = 1.6 m      _____

27. A contractor makes Lally columns by filling 20 lengths of pipe, 4 inches in diameter by 8 feet long, with concrete.  How many cubic feet of concrete are used?                                              _____

28. Find the area of the rectangle shown.

2.8 m

2 m
_____

29. What is the floor surface of a space 16 feet square?        _____

30. Find the area of a floor 14 feet square.                    _____

31. Find the area of a footing 18 inches square.                _____

32. What is the ceiling area of a room which measures 12 feet on each side?   _____

33. How much ground area is taken up by a garage which is 24 feet square?    _____

34. What is the area of a circular floor that has a diameter of 16 feet?     _____

35. How many square feet of floor space are there in a circular area that is 32 feet in diameter?                                          _____

36. A circular tower has a radius of 12 feet.  What is the floor area?       _____

37. How many square feet of floor space are there in a circular area that has a radius of 22 feet?                                        _____

38. A heavy machine requires a circular concrete base that is 13 feet in diameter. How many square feet of area are there in the base?      _____

39. How many cubic feet are contained in a footing 2′ x 2′ x 2′?            _____

40. A carpenter uses a 2-foot square piece of plywood to make a circular table top 2 feet in diameter. How much stock is wasted?
(Hint:   Waste = Area of square – Area of circle)

1′ RAD.

2′

2′
_____

# Unit 31 USING SQUARE ROOT TO FIND SIDES OF RIGHT TRIANGLES

## BASIC PRINCIPLES OF SQUARE ROOT

The *hypotenuse* of a right triangle is the side opposite the right angle. The remaining 2 sides are called the *legs* of the triangle. When the lengths of any 2 sides of a right triangle are known, the length of the third side can be found by using a rule called the Pythagorean Theorem.

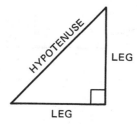

**The Pythagorean Theorem:** The square of the hypotenuse of a right triangle is equal to the sum of the squares of the 2 legs.

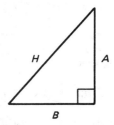

Stated as a formula: $H^2 = A^2 + B^2$ where $H$ = hypotenuse; $A$ = altitude, $B$ = base

This theorem can be illustrated by the areas of the squares in the figure:

$$H^2 = A^2 + B^2$$
$$25 = 16 + 9$$

It is also true that:

$$A^2 = H^2 - B^2$$
$$16 = 25 - 9$$

and

$$B^2 = H^2 - A^2$$
$$9 = 25 - 16$$

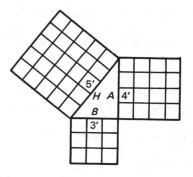

$H$ = HYPOTENUSE
$A$ = ALTITUDE
$B$ = BASE

The following are equivalent forms of this theorem:

$$H = \sqrt{A^2 + B^2} \qquad A = \sqrt{H^2 - B^2} \qquad B = \sqrt{H^2 - A^2}$$

**To find one side of a right triangle when 2 sides are known:**

1.  Choose the appropriate form of the formula.
2.  Substitute the lengths of the 2 known sides.
3.  Perform the operations.

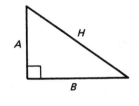

**Example:**  If $H$ = 13 cm and $B$ = 12 cm, find $A$.

1.  $A = \sqrt{H^2 - B^2}$
2.  $A = \sqrt{(13\text{ cm})^2 - (12\text{ cm})^2}$
3.  $A = \sqrt{169\text{ cm}^2 - 144\text{ cm}^2} = \sqrt{25\text{ cm}^2} = 5\text{ cm}$

Study the Framing Square method for finding the length of one side of a right triangle.

**The Framing Square Method:**  Carpenters often use the framing square to approximate the length of one side of a right triangle when the other 2 sides are known.  The scales on the outer edges of the blade and tongue are divided into 12ths on the back side of the square.  Using the back of the square, a measurement given in feet and inches can be located by reading the 12ths as inches and the inches as feet.  If measurements are given in inches and fractions of an inch, the face of the square should be used.

**To find the hypotenuse when the lengths of the 2 legs are known:**

1.  Locate the length of one leg on the outer edge of the blade.
2.  Locate the length of the second leg on the outer edge of the tongue.
3.  The distance between these two locations, measured by a carpenter's rule or a second square, is the length of the hypotenuse.

**Example:**    The 2 legs of a right triangle measure 9 feet 7 inches and 13 feet 10 inches.  Find the length of the hypotenuse.

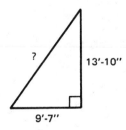

1.  On the back of the square locate $13\frac{10''}{12}$ on the body.  This will represent 13'-10''.

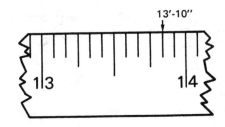

2.  On the tongue, find $9\frac{7''}{12}$. This will represent 9'-7".

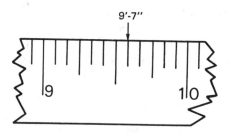

3.  The distance between these two points if read on a carpenter's rule is slightly greater than $16\frac{3}{4}$ inches. This is interpreted as slightly greater than 16 feet 9 inches.

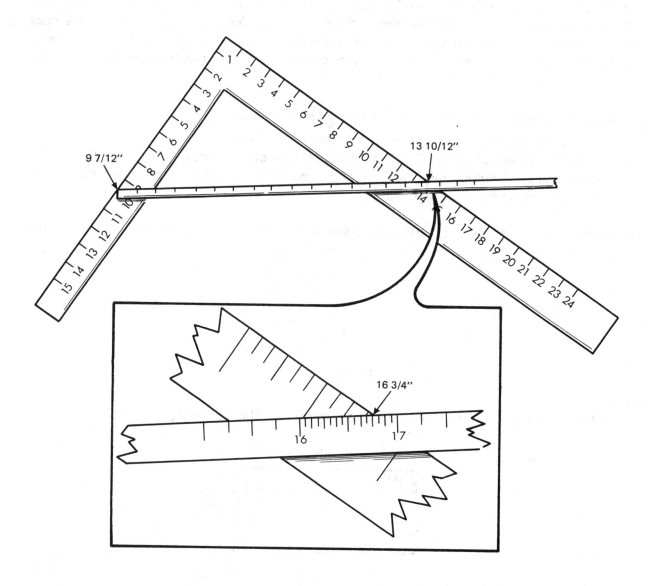

**To find the leg when the lengths of the hypotenuse and the other leg are known:**

1. Locate the known leg on the outer edge of the tongue.

2. Locate the length of the hypotenuse on the body of a second square or a carpenter's rule.

3. Place them as shown in the following figure and example.  Then, read the length of the unknown side on the body of the second square or carpenter's rule.

**Example:**  Find the missing leg of the triangle with a hypotenuse of 13 inches and a leg of 5 inches.

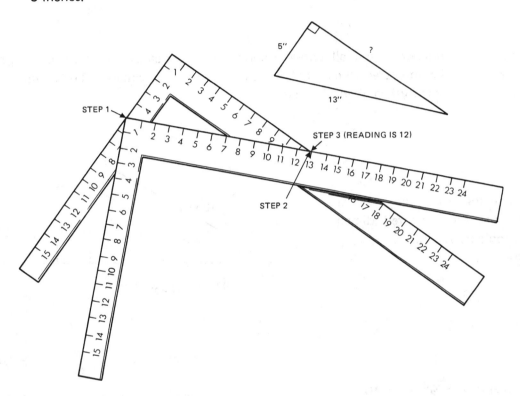

1. Locate 5 in. on the tongue of the first square and place the heel of a second square at that point.

2. Locate a hypotenuse of 13 in. on the body of the second square.

3. At the point where the 13 crosses the first square, read the length, 12 in., on the body of the first square.

## REVIEW PROBLEMS

Unless otherwise indicated, find square roots of all problems correct to the nearest hundredth. Find the following square roots:

1. $\sqrt{81}$ _____

2. $\sqrt{100}$ _____

3. $\sqrt{169}$ _____   7. $\sqrt{892}$ _____

4. $\sqrt{361}$ _____   8. $\sqrt{1,235}$ _____

5. $\sqrt{529}$ _____   9. $\sqrt{1,692}$ _____

6. $\sqrt{743}$ _____

Unless otherwise indicated, find all answers correct to the nearest 1/8th inch.  In problems 10–24, estimate the indicated dimension by using the framing square.  Then find the exact dimension by using the Pythagorean Theorem.

**Note:**  Use the right triangle shown to solve for the indicated dimension in problems 10–15.

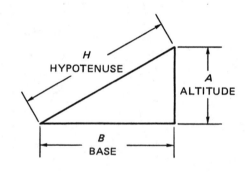

| | ESTIMATE | EXACT |
|---|---|---|
| 10.  Find $H$ if $A$ = 9″, $B$ = 10″ | _____ | _____ |
| 11.  Find $H$ if $A$ = 11″, $B$ = 13″ | _____ | _____ |
| 12.  Find $B$ if $H$ = 7″, $A$ = 5″ | _____ | _____ |
| 13.  Find $B$ if $H$ = 12″, $A$ = 8″ | _____ | _____ |
| 14.  Find $A$ if $H$ = 27 cm, $B$ = 13 cm | _____ | _____ |
| 15.  Find $A$ if $H$ = 33 m, $B$ = 18 m | _____ | _____ |

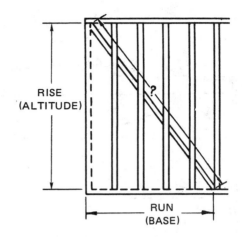

**Note:** Use the illustration of the wall brace shown for problems 16–24. In each problem find the length of the brace correct to the nearest 1/8th inch.

|  | ESTIMATE | EXACT |
|---|---|---|
| 16. Wall height = 6'-0'', Run of brace = 9'-0'' | | |
| 17. Wall height = 7'-0'', Run of brace = 11'-0'' | | |
| 18. Wall height = 6'-4'', Run of brace = 13'-0'' | | |
| 19. Wall height = 5'-2'', Run of brace = 5'-8'' | | |
| 20. Wall height = 3'-8'', Run of brace = 6'-6'' | | |
| 21. Wall height = 8'-0'', Run of brace = 13'-0'' | | |
| 22. Wall height = 18'-0'', Run of brace = 26'-0'' | | |
| 23. Wall height = 15'-2'', Run of brace = 17'-4'' | | |
| 24. Wall height = 10'-3'', Run of brace = 13'-7'' | | |

**Note:** Use the illustration of a stairway for problems 25–27. In each problem find the length of the stringer required for the stairway.

25. Total rise = 12'-0'', Total run = 16'-0''  _____

26. Total rise = 7'-6'', Total run = 5'-2''  _____

27. Total rise = 3'-9'', Total run = 2'-7''  _____

Note: Use this illustration
for problems 28–32.

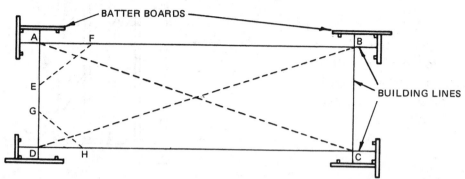

In the illustrated figure, rectangle ABCD represents the lines of excavation for the foundation of a house.

28. If line AB = 45'-0" and AD = 27'-0", what is the length of diagonals AC and BD?

_____

29. Find the lengths of the diagonals AC and BD, if the side AB is 43'-6" and AD is 32'-6".

_____

30. In checking to find out if the corner D is square, GD is laid off equal to 9'-0" and DH is laid off equal to 12'-0". What should the length of the line GH be?

_____

31. What are the lengths of diagonals AC and BD, if AB and BC are each 41'-6"?

_____

32. To square up the corner A, AE is laid off equal to 6'-0" and AF equal to 8'-0". What length is the line EF?

_____

Note: Use this illustration
for problems 33–40.

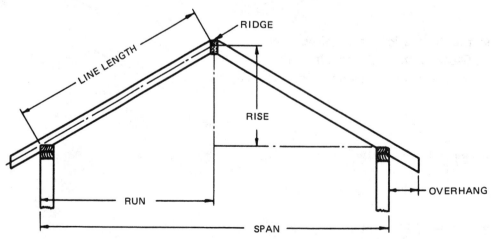

**Note:** Span, run, and overhang are terms used by carpenters for horizontal dimensions. Study the illustration and become familiar with these terms before proceeding.

For problems 33–37 find the length of the common rafters.

33. Span = 28'-0", Total rise = 7'-0", Overhang = 1'-6"  _____

34. Span = 14'-0", Total rise = 4'-8", Overhang = 1'-3"  _____

35. Span = 16'-0", Total rise = 4'-0", Overhang = 0'-8"  _____

36. Span = 46'-0", Total rise = 11'-0", Overhang = 1'-4"  _____

37. Span = 27'-0", Total rise = 9'-0", Overhang = 2'-8"  _____

38. In the illustration of the roof, what is the line length of the rafter if the run is 12'-0" and the rise is 8'-0"?  _____

39. A common gable roof, similar to the one shown, has a span of 36'-0" and a rise of 9'-0". What is the line length of the common rafter?  _____

40. Determine the line length of a common rafter for a roof that has a run of 16'-0" and a rise of 10'-8".  _____

# Estimating

## Unit 32   GIRDERS

### BASIC PRINCIPLES OF GIRDERS

*Girders* are the large beams supporting the first floor of a house.  These carry the ends of the floor joists where no wall occurs.  Girders are usually supported at the ends by the outside foundation walls, and between the walls by *Lally columns* or *posts*.

Lally columns or posts, are spaced at intervals which are determined by the ability of the beam to support the load it is designed to carry .  The size of the girders and the spacing of the Lally columns are determined by the architect and are shown on the working drawing of the basement plan.

Girder stock is purchased in lengths which are multiples of **2** feet.  All joints should be made over columns, piers, or Lally columns, the spacing of which will determine the lengths of the girders. A four-inch bearing surface under each end of a girder is usually considered a minimum.

In listing solid girders, allowance must be made for the joists and bearing as shown in the diagrams. A *built-up girder* is made of several pieces of lumber nailed together, as in the illustration.

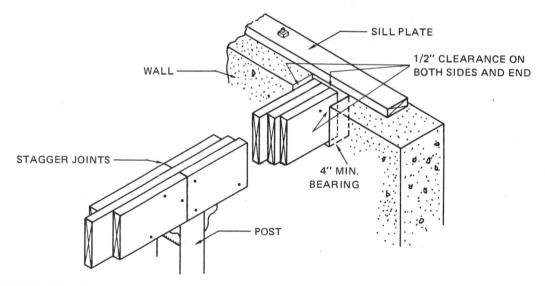

### REVIEW PROBLEMS

**Note:**   An estimate for the length of girder stock required must include waste.  Girder stock is purchased in lengths which are multiples of **2** feet.

1.   The girder span in a house is 29'-0''.  How many board feet of a solid
     6'' x 8'' girder are required?

2. A house requires two solid girders of 8″ x 8″ running parallel to each other. The span is 31′-6″. How many board feet are required? _____

3. How many board feet of material are required to construct a built-up girder (6″ x 8″) for a building 30 feet long? _____

4. From the information given, determine the number of board feet of material required for the solid girder in the illustration.

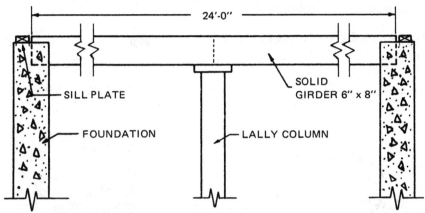

5. A built-up girder, consisting of three pieces of 2″ x 10″ stock spiked together, is needed to cover a span of 46′-0″. How many board feet are required to construct this girder? _____

6. The span of a house is 32′-6″. A chimney, 24 inches square, is located 14 feet in from one end of the girder. If the girder is a solid 6″ x 8″, how many board feet are required? _____

7. The overall width of a foundation wall is 44′-8″. The ends of the solid girder (6″ x 8″) are set in the foundation wall 6 inches in from the outside of the wall. How many board feet of stock are required for the solid girder? _____

8. A warehouse is 156 feet long with a span of 60 feet. The 6″ x 8″ girders are spaced 12′-0″ o. c. (on center). The span equals the length of the girder.

   a. How many girders are needed? _____

   b. How many board feet are required? _____

9. a. How many Lally columns are required in a building with a span of 40 feet when they are spaced 8′-0″ on center? _____

   b. What is the total cost if each Lally column costs $5.25? _____

# Unit 33   SILLS

## BASIC PRINCIPLES OF SILL USE AND MEASUREMENT

A *sill* rests on top of the foundation and is anchored to it by bolts embedded in the top of the foundation. The distance of the sill from the outside edge of the foundation is usually equal to the thickness of the wall sheathing. The sill serves as a base on which the ends of the first-floor joists rest. In some cases, the wall studs are nailed to the sill.

Standard sill stock lengths are multiples of two feet, up to 26'-0'' long. Sill lengths greater than 26 feet are called special lengths.

Study the types of sills shown. These are used in *taking-off,* or estimating, quantities in the following problems.

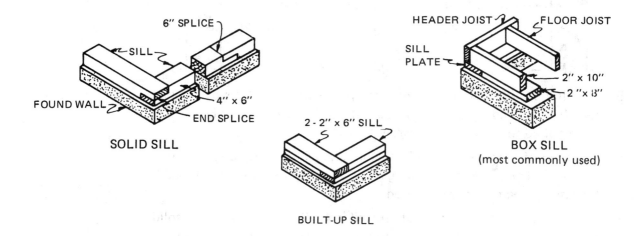

## REVIEW PROBLEMS

1. The sill shown is 2'' x 6'' x 24'-0''. How many board feet are purchased for the sill?

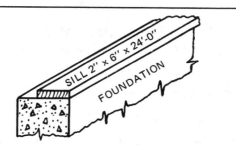

2. How many board feet are required if the sill is 4'' x 4'' x 24'-0'' long?

3. a. How many board feet of material (2'' x 6'' stock) are required to construct a 4'' x 6'' built-up sill for a house foundation 24'-0'' x 32'-0''?

   b. If 16-foot lengths are used, how many 16-foot lengths are required?

4. How many board feet of sill plate are required to complete the box sill shown? The plate is constructed of 2'' x 8'' stock. The foundation measures 28'-0'' x 36'-0''.

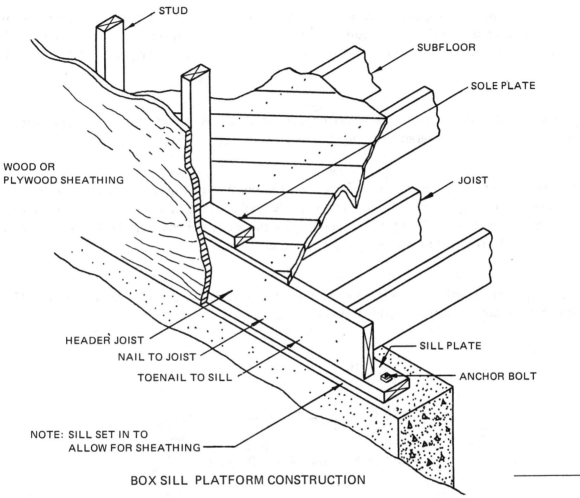

STUD

SUBFLOOR

SOLE PLATE

WOOD OR PLYWOOD SHEATHING

JOIST

HEADER JOIST

SILL PLATE

NAIL TO JOIST

TOENAIL TO SILL

ANCHOR BOLT

NOTE: SILL SET IN TO ALLOW FOR SHEATHING

BOX SILL PLATFORM CONSTRUCTION

5. How many board feet are required for the solid sill of a building 100 feet long and 40 feet wide if 4'' x 6'' stock is used?

6. How many 14-foot lengths of 4'' x 6'' stock are needed to construct a solid sill for a building 40 feet wide and 100 feet long?

7. A sill of 4'' x 6'' is installed on a foundation 40'-0'' x 32'-0''. A built-up sill 4'' x 6'' is constructed using doubled 2 x 6s.

   a. How many linear feet of stock are needed?

   b. How many board feet are needed?

 *Unit 34   FLOOR JOISTS*

## BASIC PRINCIPLES OF FLOOR JOIST CONSTRUCTION

Joists are beams used to support a floor or ceiling. Usually the size, spacing, and direction of the first-floor joists of a building are found on the foundation plan; specifications for the second-floor joists, on the first-floor plan; and specifications for the attic-floor joists, on the second-floor plan.

Floor joists run at right angles to the girder or wall which carries them. For purposes of listing or estimating, they are considered to extend to the outside face of the sills.

If the architect's drawings include a framing plan, the number of joists needed and the lengths may be counted directly from the plan. If no plan is available, the number of joists needed may be computed by dividing the length of the building by the on center spacing. To this answer, add 1 (the starter). The result is the number of joists needed.

For example, if the length of a building is 48 feet, and joist spacing is 16 inches on center (16 in. o. c.), the number of joists needed is found as follows:

$$16 \text{ in.} = \frac{4}{3} \text{ ft.}$$

$$48 \text{ ft.} \div \frac{4}{3} \text{ ft.} = 36 \text{ joists}$$

$$36 \text{ joists} + 1 \text{ starter joist} = 37 \text{ joists}$$

Standard lengths of joist stock between 8 feet and 24 feet are multiples of 2 feet.

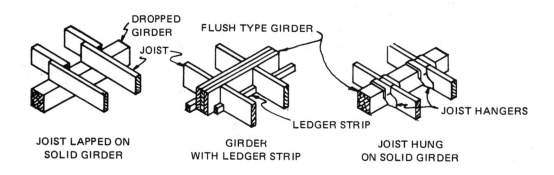

JOIST LAPPED ON SOLID GIRDER          GIRDER WITH LEDGER STRIP          JOIST HUNG ON SOLID GIRDER

## REVIEW PROBLEMS

1.  A building with a 26'-0" x 44'-0" foundation has floor joists 16 inches on center as shown. How many board feet of 2" x 8" x 14' joists are needed for the first floor?

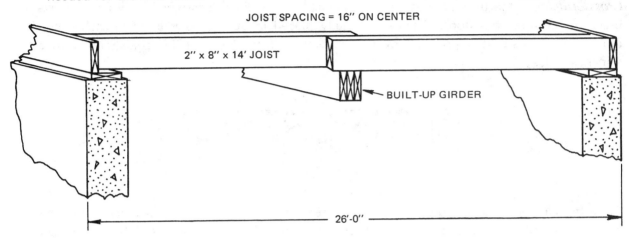

JOIST SPACING = 16" ON CENTER

2" x 8" x 14' JOIST

BUILT-UP GIRDER

26'-0"

2.  a. How many floor joists are required in a building 32 feet long x 16 feet wide if the joists are spaced 12 inches on center?

    b. If 2" x 8" x 16'-0" joists are used, how many board feet are required for the joists?

3.  A farm building is 84'-0" long, x 16'-0" wide with joists spaced 24 inches on center. If the joists are 16'-0" long, how many board feet of 2" x 8" floor joists are needed?

4.  A building 24'-0" long x 12'-0" wide has floor joists spaced 16 inches on center. Determine the board measure of the floor joists if 2" x 10" x 12'-0" joists are used.

5.  A house requires 44 pieces of 2" x 8" x 16'-0" for first floor joists as shown.

    a. How many board feet are required?

    b. What is the approximate length of the house if the joists are spaced 16 inches on center?

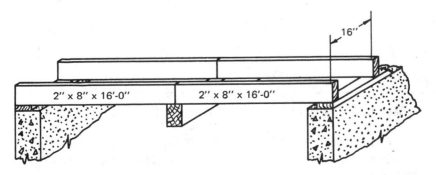

16"

2" x 8" x 16'-0"          2" x 8" x 16'-0"

# Unit 35  BRIDGING

## BASIC PRINCIPLES OF BRIDGING

*Cross bridging* is the term applied to the diagonal bracing which is fastened between the joists to brace or reinforce the floor.  They are usually placed in double rows crossing each other, as illustrated.  Cross bridging is made of material varying from 1 to 2 inches thick, and from 2 to 4 inches wide.

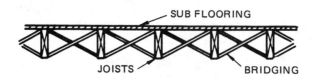

Cross bridging is nailed in double rows not more than 8'-0'' apart.  Any joist span 14'-0'' wide, or under, should have at least one double row of bridging.  A span over 14'-0'' should have two double rows of bridging.

The number of linear feet of bridging required is found by using the formula,

$$B = L \times N \times 3$$

where *B* is the number of linear feet of cross bridging, *L* is the length of the building in feet, and *N* is the number of double rows of bridging.  This formula allows for waste.

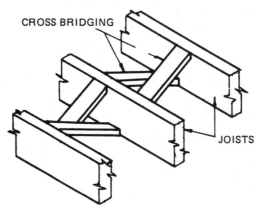

## REVIEW PROBLEMS

1.  How many linear feet of bridging are required in a building 12' x 28'?  _____

2.  A floor span is 16 feet wide and 30 feet long.  Joists, 2'' x 8'', are spaced 16 inches on center and have no center support.  How many linear feet of 1'' x 3'' stock are needed for bridging?  _____

3. A floor area 12 feet wide and 36 feet long is made up of 2″ x 8″ joists. These are spaced 16 inches on center and have no center support. How many linear feet of 1″ x 3″ stock are needed for bridging the floor?

4. A floor span 10′-0″ wide and 28′-0″ long has 2″ x 8″ joists spaced 16 inches on center. How many linear feet of 1″ x 2″ stock are needed for bridging?

5. Stock 1″ x 4″ is used for bridging a floor area 24 feet wide and 32 feet long. The joists are 2″ x 8″, spaced 16 inches on center, and are supported at their midpoint by a girder. How many board feet of stock are needed for bridging?

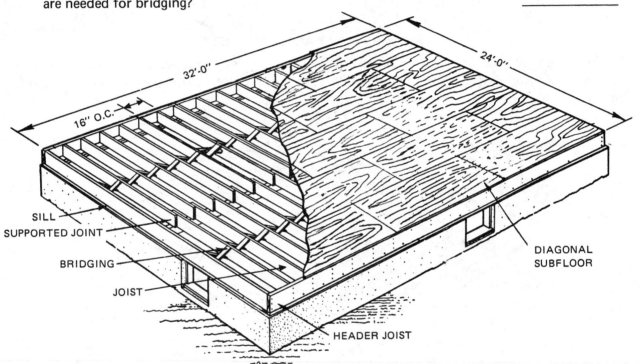

6. How many board feet of 1″ x 3″ stock are required to run two rows of bridging in a building 36 feet long?

7. Find the number of linear feet of 1″ x 4″ bridging needed in a building 24′ x 60′.

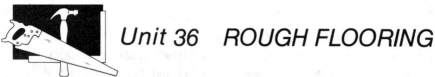

# Unit 36   ROUGH FLOORING

## BASIC PRINCIPLES OF ROUGH FLOORING

*Rough flooring,* or *subflooring,* is made up of boards or plywood laid directly on the floor joists, and upon which a finish floor is laid. Subflooring is installed either diagonally or at right angles to the floor joists.

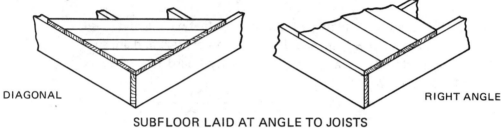

DIAGONAL

RIGHT ANGLE

SUBFLOOR LAID AT ANGLE TO JOISTS

When boards are used for subflooring, allowance must be made for waste. This allowance is usually a percentage of the area to be floored. The percentage to be added depends on the width of the boards and how they are to be laid.

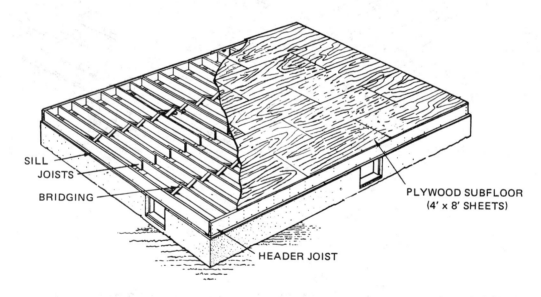

SILL
JOISTS
BRIDGING
PLYWOOD SUBFLOOR
(4' x 8' SHEETS)
HEADER JOIST

Usually, for 1'' x 6'' matched boards applied at right angles to the floor joists, 25% is added for waste and matching. For 1'' x 8'' matched boards laid at right angles to floor joists, 20% is added. Another 5% is added when rough floorboards are to be laid diagonally. If there are any openings in the floor, the area of the opening is deducted from the total area.

The use of plywood for subflooring has become increasingly popular in recent years, largely because of its ease of installation and the resulting saving in labor costs. Sold by the square foot or by the sheet, the most common sheet size is 4' x 8'. The amount of plywood needed may be estimated by the square foot. Plywood should be installed with the grain direction of the outer plies running at right angles to the joist. The sheets should be staggered so that end joists in adjacent panels break over different joists.

## REVIEW PROBLEMS

1. In the illustration, a platform step is shown. How many board feet of subflooring are needed to cover the platform if 1″ x 6″ matched boards are used?

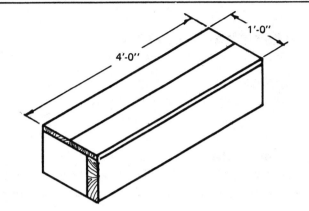

4′-0″

1′-0″

2. How many board feet of 1″ x 6″ matched rough flooring are needed for a building 32′ x 40′ if the subflooring is laid at right angles to the floor joists?

3. How many board feet of 1″ x 6″ matched boards are needed for a floor 32′ x 40′ if the subflooring is laid diagonally?  (Add 30% for waste and matching.)

4. How many board feet of 1″ x 8″ matched boards are needed to cover the floor of a building 40′-0″ x 100′-0″?  The floor is to be laid diagonally.  (Add 25% for waste and matching.)

5. A two-story building 40′ x 60′ is to have subflooring laid at right angles to the joists.  How many board feet of 1″ x 8″ matched boards are needed to cover both floors?  (Add 20% for waste and matching.)

6. How many square feet of area must be covered on a foundation 28 feet wide and 36 feet long?

7. Estimate the number of square feet of plywood needed to completely cover a floor 28′-0″ wide and 48′-0″ long.

# Unit 37   WALL PLATES

## BASIC PRINCIPLES OF WALL PLATES

*Plates* are the horizontal structural members to which the tops and bottoms of the studs are secured. The plate to which the base of the studs are nailed is sometimes called the *sole plate.*

Top and bottom plates of walls and partitions are considered separately from the studs. Plates running along the top of the studs on the exterior walls are usually made up of 2″ x 4″ lumber nailed together with a lap joint at the corners. These usually act as a seat for the second-floor joists or as a seat for the rafters in a one-story house. (See illustration.)

Bearing partition plates are made the same as the plates of exterior walls. In certain types of construction, the bearing partition studs rest on the subfloor above the girder; therefore, a single 2″ x 4″ member is used as a base to which the studs are fastened. In some cases, the outside walls also rest on top of the subfloor and would also require a sole plate.

No allowance is made for openings when estimating the linear footage of plates. Express the linear footage into the number of standard lengths required by dividing the standard length into the total linear footage.

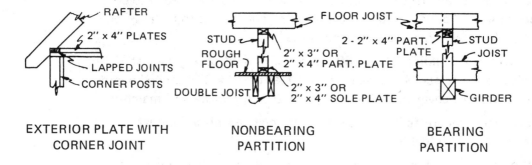

EXTERIOR PLATE WITH CORNER JOINT     NONBEARING PARTITION     BEARING PARTITION

## REVIEW PROBLEMS

1. How many board feet of 2″ x 4″ stock are needed to construct the sole and top plates shown?

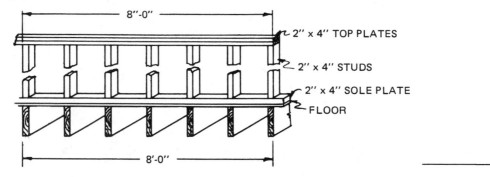

2. How many board feet of single sole and double top plate of 2″ x 4″ stock are needed for the plan shown?

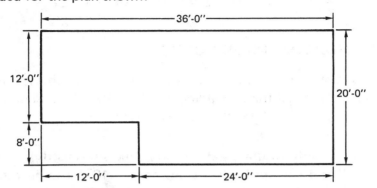

3. How many board feet of single sole and double top plate of 2″ x 4″ stock are needed for the construction of the outside walls of a building which measures 34'-0″ x 46'-0″?

4. Estimate the number of board feet of single sole and double top plate of 2″ x 4″ stock needed for the building shown.

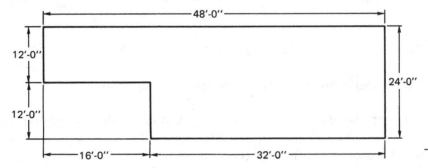

5. Find the number of 2″ x 4″ x 16'-0″ lengths required for the outside double plates of a house 24' x 42'.

6. How many 2″ x 4″ x 16'-0″ lengths are required for the exterior double plates of a house that is 32'-0″ x 48'-0″?

# Unit 38  STUDDING AND FIRE OR DRAFT STOPS

## BASIC PRINCIPLES FOR STUDS AND FIRE STOPS

*Studs* are the vertical members of outside walls or partitions to which sheathing, panelling, or laths are fastened. Studs for small frame buildings are 2'' x 4'' or less. Some contractors use 2'' x 6'' on outside walls and 2'' x 4'' or 2'' x 3'' on interior partitions.

A common practice, and a fairly accurate way of estimating studs used, is to allow one stud for every linear foot of wall and partition, when studs are set 16 inches on center. This surplus allows for the doubling at corners and around openings. Studs may be obtained in multiples of 2'-0'' in length, or may be purchased precut.

To estimate the number of studs required in a building, the following steps are usually followed:

1.  Calculate the perimeter of a building to determine the number of studs needed for the outside walls.

2.  Determine the number of linear feet of all bearing partitions.

3.  Determine the number of linear feet of all nonbearing partitions.

4.  Total the perimeter and the number of linear feet of all partitions, bearing and nonbearing.

5.  Allow one stud for every linear foot of the total.

A *draft stop,* or *fire stop,* is material inserted in the space between studs or joists to prevent the spread of fire or smoke. It is placed so that it will block the passage of flames or air currents upward or across the building.

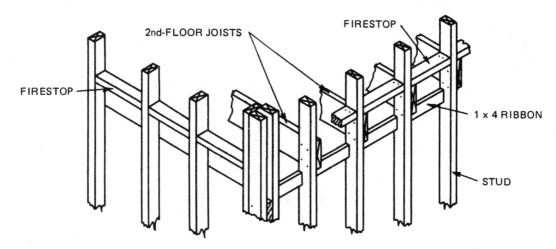

## REVIEW PROBLEMS

1. In the illustration, the fire stops occur between the joists at the second-floor level. How many linear feet of joist material are needed for the stops?

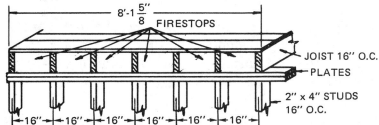

2. A house is 24'-0'' wide and 32'-0'' long. The studs are spaced 16 inches on center. At the first-floor level, how many linear feet of draft stops should be used around the outside walls.

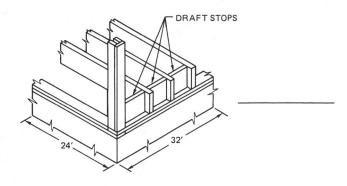

3. How many outside studs, spaced 16 inches on center, are required to construct the outside walls of a building that is 26'-0'' wide and 38'-0'' long?

4. How many board feet of 2'' x 4'' stock are required for the studs in the outside walls of a building 26'-0'' x 38'-0'' if the studs are 8'-0'' long and spaced 16 inches on center?

5. How many linear feet of outside studding 8 feet long are required for a house 24'-0'' x 42'-0''? Studs are 16 inches on center.

6. Estimate how many board feet of studding are needed for a house with the following specifications: (All studs are spaced 16 inches on center.)

   a. Overall measurements are 24'-0'' wide x 32'-0'' long. (All studs are 2'' x 4'' x 10'-0'')  a _____

   b. One bearing partition, 32'-0'' long; one, 14'-0'' long. (All studs are 2'' x 4'' x 8'-0'')  b _____

   c. Three nonbearing partitions, 10'-0'' long (All studs are 2'' x 3'' x 8'-0'')  c _____

   d. One nonbearing partition, 6'-0'' long. (All studs are 2'' x 3'' x 8'-0'')  d _____

   e. Estimate the total number of board feet of studs required for the outside walls and partitions found in a through d.  e _____

7. A house has 120 linear feet of outside walls, 64 linear feet of bearing partitions, and 130 linear feet of nonbearing partitions. The length of all studs is 8'-0''. Estimate the number of 16'-0'' lengths that are required for all the studding.

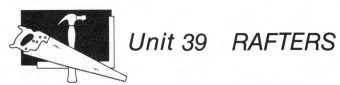

# Unit 39  RAFTERS

## BASIC PRINCIPLES FOR RAFTERS

*Rafters* are the main structural roof supports.  The *ridge* of a roof is the uppermost structural member of a roof.  The *top plate* is the horizontal structural member on top of the wall studs. *Common rafters* run from the plate to the ridge.  A *hip rafter* is a rafter at the intersection of two slopes of a roof which form an external angle.

Dimensions used in calculating rafter lengths are illustrated.  *Line length* is the hypotenuse of a right triangle the base of which is the run, and the height of which is the rise.  *Overhang* is the horizontal distance the rafter projects beyond the wall on which it rests.  The *tail* is the distance along the center line of the rafter that extends beyond the wall.

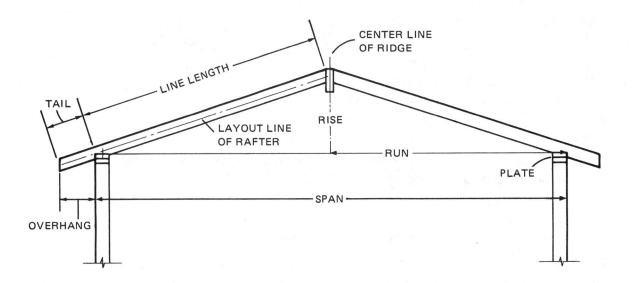

Carpenters normally use the rafter table on the steel square to determine the rafter length.  In special cases the carpenter might use the Pythagorean Theorem.  Rafter stock is sold in lengths from 8 feet to 24 feet in multiples of 2 feet. When an estimate of the number of rafters needed for a common gable roof is made, one rafter should be added to the total for each side of the roof.

**Example:**  What is the length of the rafters of a gable roof which has a 1/3 pitch (8 inch rise per foot of run).  The roof span is 24'-0'' and the overhang is 1'-0''.

Solution:  Rafter run = 24 feet ÷ 2 = 12 feet
Rafter run and overhang = 12 feet + 1 foot = 13 feet
(Using the 8-inch rise on the steel square, the length of the common rafter is 14.42 inches per foot of run.)

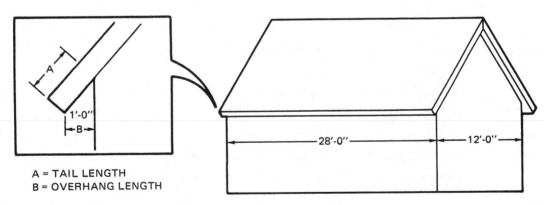

Rafter length = 13 x 14.42 inches = 187.46 in.
                = 15 feet 7 15/32 inches

## REVIEW PROBLEMS

**Note:** Rafter material is purchased in 2 foot multiples. Use these rafter tables on the following pages to solve 1–7.

1. The roof illustrated has a rise of 6 inches for every foot of run and an overhang of 1 foot.

   a. Find the length of the common rafters.
   
   a. _____

   b. Determine the number of board feet of stock required for all of the common rafters in the roof. Stock used for rafters is 2" x 6". (16 in. o. c.)
   
   b. _____

A = TAIL LENGTH
B = OVERHANG LENGTH

2. The gable roof of a house, having a span of 24 feet and length of 36 feet, has a rise of 8 inches per foot of run.  The rafters are 2" x 6", spaced 16 inches on center, and have a 12-inch overhang.

   a. Find the length of the common rafters.   _____

   b. Determine the total number of common rafters needed for the roof.   _____

   c. Calculate the number of board feet of stock required for the common rafters.   _____

3. Find the number of board feet of stock required for the rafters of a gable roof which has a 12-inch rise per foot of run, a span of 26'-0", and a roof length of 44'-0".  Use 2" x 6" rafters spaced 16" on center and allow for an overhang of 1'-0".   _____

Note:  Use this illustration for problems 4—7.

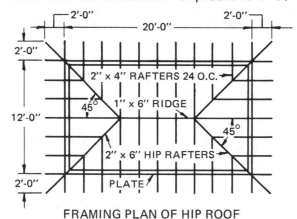

FRAMING PLAN OF HIP ROOF

4. Find the length of the hip rafter if the roof shown has a rise of 8 inches for each foot of run.   _____

5. Find the length of all the hip rafters to be ordered for the roof shown if it has a 6 inch rise for each foot of run.   _____

6. What is the length of all the hip rafters needed for the roof shown if the roof shown has a rise of 9 inches for each foot of run?   _____

7. What is the total length of the hip rafters shown, if the roof has a rise of 4 inches for each foot of run?   _____

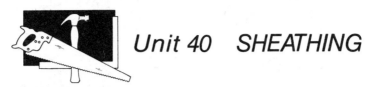

# Unit 40   SHEATHING

## BASIC PRINCIPLES OF SHEATHING

*Sheathing* is the structural covering over the exterior side of joists, studs, or rafters. The most commonly used materials are boards (applied diagonally or horizontally), plywood, and structural insulation board.

Sheet materials, such as plywood or structural insulation board, are the most common sheathing materials used today. Sheet materials are sold by the square foot, commonly in sizes either 2' x 8' or 4' x 8' sheets. The 4' x 8' sheets are most readily available. The amount required is estimated by the number of square feet to be covered.

Board sheathing is still sometimes used. An estimate of the amount required is computed by determining the number of square feet of wall area to be covered. From this amount, deduct the area of all openings (such as doors and windows) and add a percentage for waste (25% of whole area for horizontal or 30% of whole area for diagonal). The result will indicate the number of square feet of material to be ordered.

The amount of roof boards required is also calculated on the basis of the square footage of roof area to be covered. If 1'' x 6'' or 1'' x 8'' boards are used, an allowance of 20% is added to compensate for waste.

The illustrations which follow indicate the usual methods of applying sheathing materials.

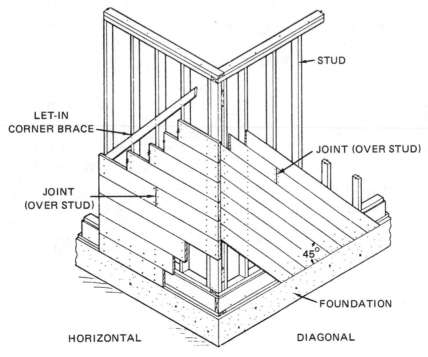

INSTALLATION OF BOARD SHEATHING

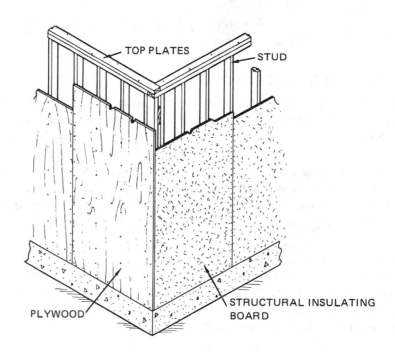

TOP PLATES

STUD

STRUCTURAL INSULATING BOARD

PLYWOOD

INSTALLATION OF PLYWOOD OR STRUCTURAL INSULATING BOARD SHEATHING

## REVIEW PROBLEMS

1. How many square feet of 1'' x 8'' board sheathing are required to diagonally sheath a wall 8' x 26'?  Make no allowance for openings.    _____

2. A building measuring 8' x 26' x 42' is to be sheathed using horizontal 1'' x 8'' boards.  Allowing a total of 187 square feet for openings, how many square feet of sheathing are required?    _____

3. Plywood is used as sheathing for the exterior walls of a home which measures 8' x 28' x 48'. How many 4' x 8' sheets of 1/2-inch plywood are needed?    _____

4. A wall 37 feet long is to be sheathed to a height of 12 feet.  How many square feet of structural insulation board 25/32-inch thick are required?    _____

5. A gable roof is to be covered with 1'' x 8'' roof boards.  If each side measures 20' x 60', what is the total number of square feet of board, including an amount for waste, which must be ordered?    _____

6. How many 4' x 8' sheets of plywood sheathing are required for a shed roof measuring 28'-0'' x 16'-0''?    _____

# Unit 41 TRIM

## BASIC PRINCIPLES OF TRIM

*Trim* is finish material applied around the openings, ceilings, and floors of a building. Trim required for a house usually falls into two catagories, exterior and interior. Most generally, trim is available in standard lengths and is sold by the linear foot.

*Exterior trim* is that portion of the exterior finish of a house other than the wall covering. Exterior trim typically includes window and door trim, as well as those items making up the cornice.

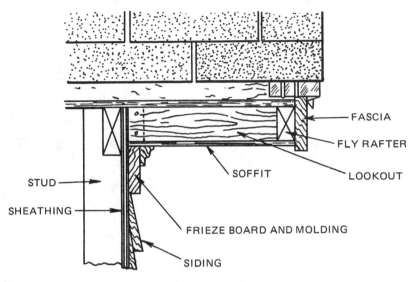

CORNICE INSTALLATION — GABLE END

*Interior trim* includes items such as door trim (jamb, stops, and casing), window trim (casing, apron, stool, stops), base trim (baseboard, base shoe, base cap), ceiling trim, and other items such as chair rail or picture moldings. In all cases, the estimator must find the number of linear feet of trim required, usually by determining the perimeter of the door, window, or room to be trimmed.

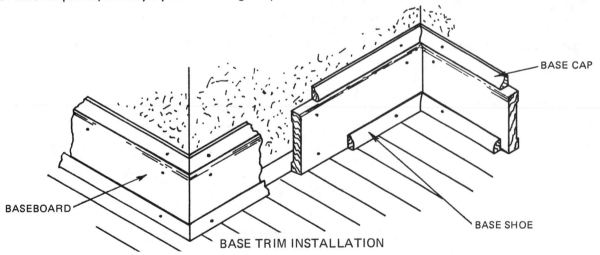

BASE TRIM INSTALLATION

## REVIEW PROBLEMS

1. Find the number of feet of frieze board required for a gable end cornice if the span is 24 feet and the roof pitch is 1/4. Express the answer to the nearest whole foot.

2. What is the actual number of feet of baseboard required for a room measuring 12' x 22'? Deduct 8 feet for openings.      _____

3. How many feet of window casing are required for two windows, each measuring 3 feet by 4 feet 6 inches? Allow 1 foot per window for waste. Do not allow for the stool or apron.      _____

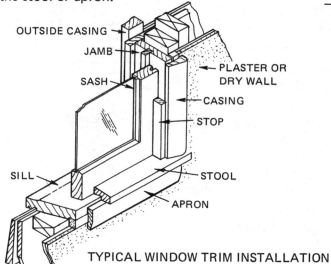

TYPICAL WINDOW TRIM INSTALLATION

4. What amount of door casing is required to trim a passage door measuring 2'-6'' x 6'-8''? Both sides of passage doors require casing.

_____

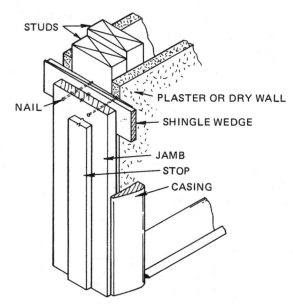

INSTALLATION OF DOOR JAMB AND CASING

# Unit 42   ROOFING

## BASIC PRINCIPLES OF ROOFING

*Roofing* is the exterior waterproof cover of a roof structure. The most common coverings for pitched roofs are asphalt, asbestos, and wood shingles. Other types of roofing such as tile, slate, metal, or built-up roofing are also used, but these types are not usually installed by the carpenter.

Wood shingles are commonly bought by the square (100 square feet). To estimate the number of squares, calculate the total number of square feet of roof surface to be covered and divide by 100 sq. ft. For plain roofs, add an allowance of 8% of the area for waste. A square of 18-inch shingles, laid 6 inches to the weather, covers 100 square feet. For other exposures, a proportional amount must be figured (at 5-inch exposure, only 5/6 as much will be covered, therefore, the total estimate must be increased by 1/5).

Asphalt or asbestos shingles are also estimated on the basis of the square (100 square feet). Again, the total roof area is divided by 100 sq. ft. to give the number of squares which must be ordered. A figure of 5% added for waste is often used.

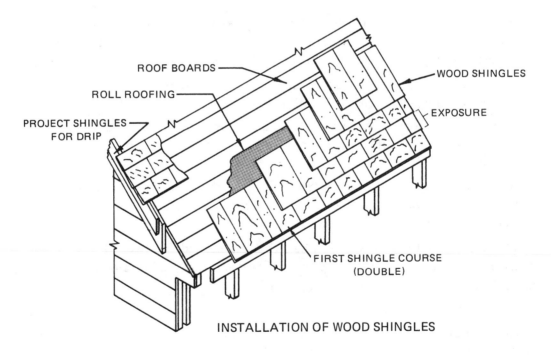

INSTALLATION OF WOOD SHINGLES

## REVIEW PROBLEMS

1.  Determine the number of bundles of asphalt shingles required for both sides of a gable roof house with a span of 26 feet (including overhang), a ridge of 40 feet, and 1/4 pitch. They are to be laid 5 inches to the weather. Each bundle contains 1/3 square. No allowance is to be made for waste.

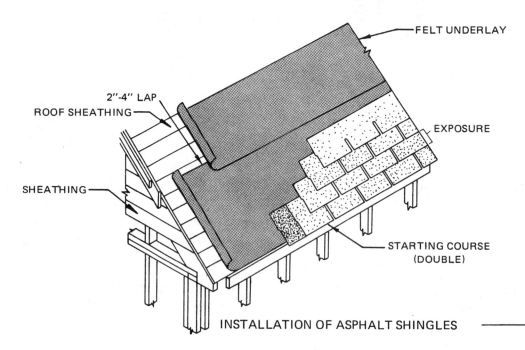

FELT UNDERLAY

2''-4'' LAP

ROOF SHEATHING

EXPOSURE

SHEATHING

STARTING COURSE
(DOUBLE)

INSTALLATION OF ASPHALT SHINGLES    _____

2. How many bundles of 18-inch wood shingles laid 6 inches to the weather are required for one side of a gable roof measuring 16 feet x 32 feet? There are four bundles per square. (Allow 8% for waste.)    _____

3. If asphalt shingles that weigh 235 pounds to the square are used, find the weight of shingles for an area of 10' x 15'.    _____

4. a. How many squares of asphalt shingles does it take for 2 sides of roof with a rafter length of 18 feet and a ridge 28 feet long? Allow 5% for waste.    _____

   b. If asphalt shingles that weigh 235 pounds to the square are used, find the total weight of the shingles used. Do not consider the weight of the 5% waste allowance.    _____

5. Find the number of squares of asphalt shingles needed for two sides of a gable roof with a 1/3 pitch, a span of 22 feet and a ridge 30 feet long.    _____

6. How many courses of shingles are required if the rafter length is 16 feet and they are laid 4 3/4 inches to the weather?    _____

7. A roll of roofing paper covers an area of 100 square feet. How many rolls are needed for a lean-to roof 12' x 16'?    _____

 # Unit 43 DOORS AND WINDOWS

## BASIC PRINCIPLES OF DOORS AND WINDOWS

Windows, exterior doors, and their frames are items which are usually made at a factory and installed, ready for use, at the construction site. Most architectural plans contain specification sheets or a door and window schedule from which the carpenter may simply count and order the specified types.

The most common window types used in residential construction are the double hung, the casement, the hopper, the awning, the fixed, and the horizontal sliding. They may be made of wood or metal and usually can be ordered with regular or insulated glass and with premade screens or storm window units.

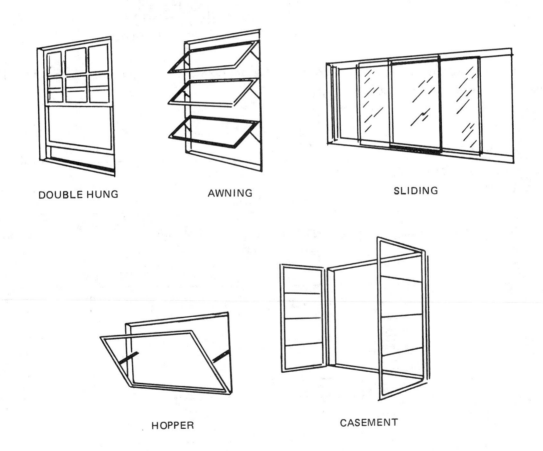

DOUBLE HUNG     AWNING     SLIDING

HOPPER     CASEMENT

Interior doors and door frames are often purchased as completed units ready for installation. However, they are often hung by the carpenter who installs the jambs, stops, and casing. The carpenter usually frames the rough opening about 3 inches higher than the door height and about 2 1/2 inches more than the door width in order to provide for jambs, and leveling within the

EXTERIOR DOORS

FLUSH    PANELED    PANELED    FLUSH    COMBINATION

INTERIOR DOORS

PANELED    FLUSH    LOUVERED-FOLDING

opening.  The most common types of doors are the paneled and the flush door.  Folding doors often used for closets are flush, louvered, or paneled.

## REVIEW PROBLEMS

**Note:** Use this chart for problems 1 and 2.

| WINDOW SCHEDULE | | | | | | | |
|---|---|---|---|---|---|---|---|
| Symbol | Item | Quantity | Type Window | Rough Opening | Unit Size | Manufacturer No. | Remarks |
| A | 1 | 1 | Bay | $9'\text{-}9\frac{1}{2}''$ x 5'-4'' | $9'\text{-}6\frac{1}{4}''$ x $4'\text{-}11\frac{3}{8}''$ | Jones 1580 | Radius 10'-4 1/2'' |
| B | 2 | 4 | Sliding | $3'\text{-}7\frac{1}{4}''$ x $3'\text{-}6\frac{1}{4}''$ | $3'\text{-}5\frac{1}{2}''$ x $3'\text{-}3\frac{1}{2}''$ | Smith 16285 | Removable Sash |
| C | 3 | 1 | Double Hung | $2'\text{-}10\frac{1}{8}''$ x $3'\text{-}1\frac{1}{8}''$ | 2'-8'' x 2'-10'' | Jones 1437 | Insulating Glass |
| D | 4 | 6 | Double Hung | $3'\text{-}2\frac{1}{8}''$ x $4'\text{-}5\frac{1}{8}''$ | 3'-0'' x 4'-2'' | Jones 1467 | Insulating Glass |
| E | 5 | 2 | Casement | 5'-4'' x 1'-8'' | 4'-11'' x 1'-3'' | Smith 782A | Insulating Glass |

1. From the window schedule illustrated, determine the total cost of item 4, if one unit costs $44.96.    _____

2. According to the window schedule illustrated, how many casement windows must be ordered?    _____

**Note:** Use this chart for problems 3—5.

| DOOR SCHEDULE | | | | | | |
|---|---|---|---|---|---|---|
| Symbol | Quantity | Type Door | Rough Opening | Unit Size | Manufacturer No. | Remarks |
| A | 1 | Panel | $3'\text{-}3\frac{1}{2}'' \times 6'\text{-}11\frac{1}{2}''$ | $3'\text{-}0'' \times 6'\text{-}8''$ | Miller - 821P | Exterior |
| B | 5 | Flush | $2'\text{-}9\frac{1}{8}'' \times 6'\text{-}11\frac{1}{8}''$ | $2'\text{-}6'' \times 6'\text{-}8''$ | Jones - 1175 | Passage |
| C | 2 | Panel | $2'\text{-}3\frac{1}{8}'' \times 6' \times 11\frac{1}{4}''$ | $2'\text{-}0'' \times 6'\text{-}8''$ | Miller - 8501 | Closet |
| D | 1 | Combination | $3'\text{-}3\frac{1}{2}'' \times 6' \times 11\frac{1}{2}''$ | $3'\text{-}0'' \times 6'\text{-}8''$ | Smith - 2750 | Exterior |
| E | 2 | Folding | $5'\text{-}3\frac{1}{4}'' \times 6' \times 11\frac{1}{2}''$ | $5'\text{-}0'' \times 6'\text{-}8''$ | Miller - 875F | Storage |
| F | 2 | Sliding | $4'\text{-}3\frac{1}{2}'' \times 6' \times 11\frac{1}{2}''$ | $4'\text{-}0'' \times 6'\text{-}8''$ | Jones - 1190 | Closet |

3. The total cost for item B is $40.55. Using the door schedule shown, find the cost of one flush-type door. _____

4. What size rough opening is required for an interior door measuring 2'-6'' x 6'-8''? _____

5. Using the door schedule, find the total cost of sliding closet doors at a unit cost of $17.51. _____

# Unit 44   SIDING

## BASIC PRINCIPLES OF SIDING

*Siding* is the exterior wall finish applied to a building. Clapboards and other types of siding are sold by the square foot. The surface area to be covered, plus a percentage for lapping and waste is calculated to determine the number of square feet required for an area.

The various types of siding shown below are usually made of 1'' x 6'' or 1'' x 8'' boards. Required quantities are computed on the same basis as sheathing, although an allowance must be made for both lap and waste. For the so-called 1'' x 6'' novelty siding, the lap and waste allowance is 25%. Builder's felt is often applied to the sheathing before the siding is put on. When estimating felt, 10% is added to the area to be covered. A roll of felt usually contains 400 square feet.

The various types of novelty siding shown below are usually made of 1'' x 6'' boards, and required quantities are computed the same as for sheathing. Celotex or another type of insulating board is sometimes applied between the sheathing and the siding. This material comes in rectangular sheets, 4'-0'' wide and in even lengths from 8'-0'' to 12'-0''.

### TYPES OF SIDING

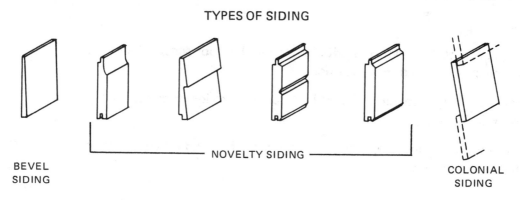

BEVEL
SIDING

— NOVELTY SIDING —

COLONIAL
SIDING

The table below gives common allowances for lap and waste on bevel siding.

| TABLE OF LAP AND WASTE ALLOWANCES — BEVEL SIDING | | |
|---|---|---|
| **Width** | **Exposure to Weather** | **Add for Lap and Waste** |
| 12'' | 10 1/2'' | 15% |
| 12'' | 10'' | 20% |
| 10'' | 8 1/2'' | 18% |
| 10'' | 8'' | 24% |
| 8'' | 6 1/2'' | 23% |
| 8'' | 6'' | 30% |
| 6'' | 5'' | 25% |
| 6'' | 4 3/4'' | 32% |
| 6'' | 4 1/2'' | 38% |
| 4'' | 2 3/4'' | 51% |
| 4'' | 2 1/2'' | 65% |

## REVIEW PROBLEMS

1.  If applied as shown below, how many square feet of 6-inch bevel siding, are required to cover the side of a house 24' x 28' x 18'-6''? Exposure to the weather is 4 1/2 inches.

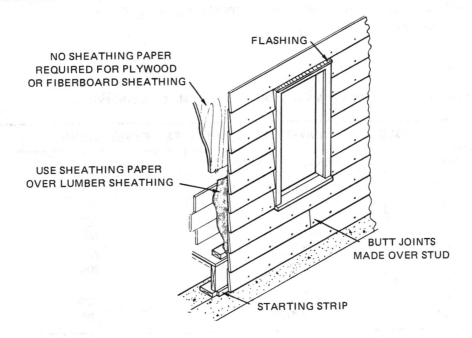

FLASHING

NO SHEATHING PAPER
REQUIRED FOR PLYWOOD
OR FIBERBOARD SHEATHING

USE SHEATHING PAPER
OVER LUMBER SHEATHING

BUTT JOINTS
MADE OVER STUD

STARTING STRIP

INSTALLATION OF BEVEL SIDING

2.  Determine the number of square feet of 8-inch bevel siding needed to cover the sides of a house 28' x 32' x 18' if the exposure to the weather is 6 1/2 inches?

3.  For a house 26' x 40' x 17'-6'', estimate the number of square feet of 10-inch bevel siding needed. Exposure to the weather is 8 1/2 inches.

4.  What is the number of square feet of 8-inch bevel siding needed to cover the sides of a house 28' x 42' x 10'? Exposure to the weather is 6 1/2 inches?

5.  A garage is to be covered with bevel siding. The two sides measure 22' x 9', and the one end, 20' x 9'. Estimate the number of square feet of 6-inch bevel siding needed if the exposure is 5 inches to the weather.

6.  How many square feet of 8-inch bevel siding, laid 6 3/4 inches to the weather, are needed for a house 26' x 40' x 12'? (Allow 22% for waste.)

7. Find the number of square feet of drop siding required for two gable ends of a building with an 18-foot span and 1/3 pitch. (Allow 25% for waste.)

_____

8. Determine the number of rolls (400 square feet per roll) of builder's felt needed to cover the walls of a house measuring 28' x 46' x 8'. (Allow 10% for waste.)

_____

The table below gives common allowances for lap and waste on bevel siding.

| TABLE OF LAP AND WASTE ALLOWANCES — BEVEL SIDING | | |
|:---:|:---:|:---:|
| **Width** | **Exposure to Weather** | **Add for Lap and Waste** |
| 12" | 10 1/2" | 15% |
| 12" | 10" | 20% |
| 10" | 8 1/2" | 18% |
| 10" | 8" | 24% |
| 8" | 6 1/2" | 23% |
| 8" | 6" | 30% |
| 6" | 5" | 25% |
| 6" | 4 3/4" | 32% |
| 6" | 4 1/2" | 38% |

# Unit 45   STAIRS AND INTERIOR DOOR JAMBS

## BASIC PRINCIPLES OF STAIRS AND INTERIOR DOOR JAMBS

The specifications which accompany each set of plans provide the necessary information on types and kinds of material which are to be used. The specifications also include detailed information on all standard or prefabricated parts. Examination of the plans provides any additional necessary information on special sizes.

Door jambs are ordered according to the number required; the kind of material; the width, thickness, and type of jamb; and the door size (width, height, and thickness).

**Example** of method of listing door jambs:

| No. | Material | Size of Jambs | Type | Door Size |
|-----|----------|---------------|------|-----------|
| 6 | N.C. Pine | 1 1/8'' x 5 1/2'' | Rabbetted | 2'-6'' x 6'-6'' x 1 3/8'' |

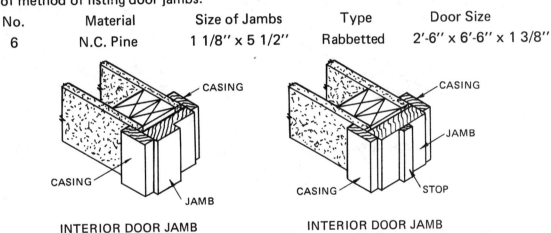

INTERIOR DOOR JAMB
RABBETTED

INTERIOR DOOR JAMB
WITH STOPS PLANTED ON

Stair material specifications must include several important items. These are the sizes of risers, treads, stringers, rails, and newel posts. Normally the sum of the measurements of tread and riser is 17 1/2 inches.

**Example:** What is the size of each tread and riser for the stairs shown?

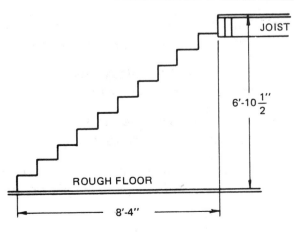

**Solution:**

1. Riser Height = 6'-10 1/2'' ÷ 11 risers
   = 82.5'' ÷ 11
   = 7.5'' or 7 1/2''
2. Tread Run = 8'-4'' ÷ 10
   = 100'' ÷ 10
   = 10''

## REVIEW PROBLEMS

1. How many risers are required for a set of stairs if the total height is 6'-6 3/8'' and the unit rise is 7 1/8 inches?

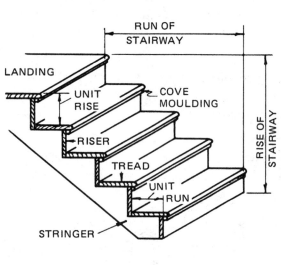

STAIR DETAIL

PARTITION DETAIL

_____

2. The unit of rise of a set of stairs is 7 inches. How many risers are needed to attain a height of 8'-2''?

_____

3. Find the unit rise and unit run of a stair to fit an opening that has a total run of 9 feet 7 1/2 inches and a rise of 7 feet. Eleven treads are used.

_____

4. Find the number of treads for a flight of stairs that has a total rise of 105 inches and a unit rise of 7 inches.

_____

5. How many 7 9/16-inch risers are there in a flight of stairs when the total rise of the stairs is 10'-1''?

_____

6. How many linear feet of 3/4'' x 4 1/4'' are required for the header and side jambs of four doors 2'-6'' x 6'-8''?

_____

7. What width door jamb is needed to frame an opening with studs, 1 1/2'' x 3 1/2'' and 1/2-inch dry wall on each side of the studs?

_____

# Unit 46   FINISH FLOORING AND PAPER

## BASIC PRINCIPLES FOR FINISH FLOORING AND PAPER

*Finish flooring* is the floor covering material which is applied as the final wearing surface of a floor. Today, many materials such as wood strips, wood blocks, a variety of tile and carpeting are used for finish floors. The types of finish flooring commonly installed by the carpenter are either of the wood strip or wood block variety. To estimate the required quantity of finish floor, compute the area of the surface to be covered. To this figure, add an allowance for waste and matching. While the specific type of flooring generally determines the exact amount, an allowance of 40% is usually considered to be sufficient.

Building paper of various kinds is used between the rough and finish flooring. The purpose of the paper is to stop dust or drafts and, to some extent, to insulate against temperature and sound. Rosin-sized paper and damp-proof papers are commonly used for this purpose. If a greater degree of insulation is desired, there are many grades of deadening felt or quilt on the market.

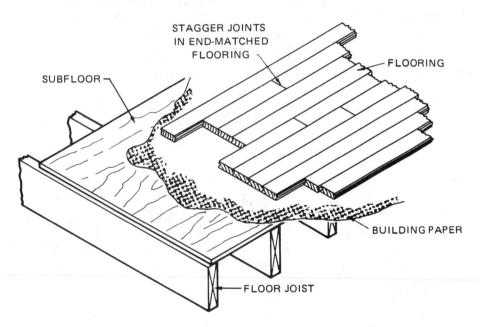

GENERAL APPLICATION OF STRIP FLOORING

The examples which follow illustrate the calculation of finish flooring and building paper.

**Example:** How many square feet of strip flooring are needed to cover a first floor plan which measures 25'-0" x 32'-0"?

**Solution:**

1. Total area = 25' x 32' = 800 square feet
2. Quantity needed = 800 + (40% x 800) = 800 + 320 = 1,120 square feet

**Example:** How many rolls of building paper are needed under this amount of finish flooring? (250 square feet in each roll)

**Solution:**

1. Total area = 25' x 32' = 800 square feet
2. Number of rolls = 800 ÷ 250 = 3 1/5 rolls
3. Number of rolls needed = 4

## REVIEW PROBLEMS

1. How many board feet of strip maple flooring are needed for a building which measures 30'-6" x 42'?

   _____

2. Find the number of board feet of flooring ordered for a floor measuring 12'-3" x 16'.

   _____

3. Find the total number of board feet of oak strip flooring necessary for the following room sizes: 9'-6" x 10'; 10'-9" x 12'-6"; 12' x 21'-6"?

   _____

4. Find the number of board feet of flooring needed for a floor laid block style if the floor measures 20'-6" x 26'-3".

   _____

5. Find the total number of square feet of flooring necessary for the following size rooms: bathroom, 9'-6" x 6'; kitchen, 11'-6" x 8'; 2 bedrooms, each 12'-3" x 14'; and a living room, 14' x 20'?

   _____

6. Estimate the number of rolls of building paper, each containing 250 square feet, needed to cover two floors, each measuring 25' x 35'.

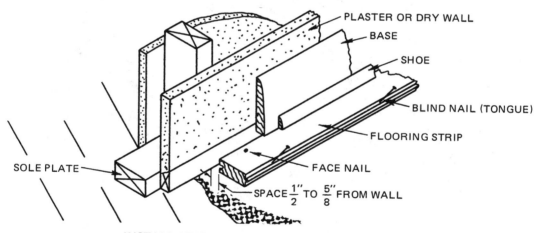

INSTALLATION OF STRIP FLOORING AT WALL

   _____

# Unit 47   HARDWARE AND SUPPLIES

## BASIC PRINCIPLES FOR HARDWARE AND SUPPLIES

The carpenter often estimates quantities of fasteners, hardware, and other supplies required in the construction of a building. Included in these categories are nails, screws, bolts, hinges, hangers, shelf and closet brackets, and adhesives. Many of these items may be estimated through a detailed study of the plans. Others are estimated on the basis of understanding gained through experience in working with these special materials.

## REVIEW PROBLEMS

1.  Using the chart, find how many pounds of 8d casing nails are required to lay 2,500 board feet of strip flooring?

| QUANTITIES OF NAILS REQUIRED | | | |
|---|---|---|---|
| **Material to be Fastened** | **Size and Kind** | **Pounds Needed** | **Unit** |
| Joists and Sills | 16d common | 20 | 1000 board feet |
| Studs, Rafters | 8d, 16d common | 20, 25 | 1000 board feet |
| Composition Shingles | 1 1/2" galvanized | 3 | square |
| Sheathing | 8d common | 20 | 1000 board feet |
| Siding | 6d common | 18 | 1000 board feet |
| Flooring | 8d casing | 30 | 1000 board feet |
| Wood Shingles | 3d galvanized | 5 3/4 | square |

_____

2.  At a rate of 3 pounds of roofing nails per square, how many pounds of nails are required for a gable roof measuring 20'-0" x 35'-0" on each side?

_____

3.  A foundation measures 26'-0" x 42'-0". Anchor bolts are placed 1 foot from each end and then every 8 feet as illustrated. How many anchor bolts are needed to place the sill?

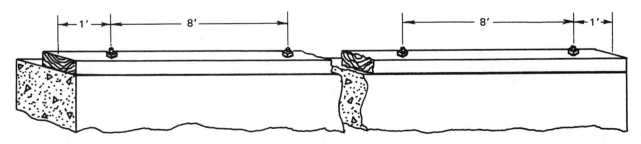

_____

4. Using one pair per door, how many butt hinges are required to hang 11 interior doors?

_____

5. Contact cement has a coverage rate of 50 square feet per gallon. How much is needed for a job measuring 12'-0" x 23'-0"?

_____

# ACHIEVEMENT REVIEW A

The following testing material is provided for the convenience of the instructor. Delmar Publishers gives permission to reproduce this material in whole or in part to meet the individual needs of the instructor.

**Note:** The numbers in parentheses, ( ), given below each question show the unit or units in which similar problems have been discussed.

1. The lower level of a house has a recreation room, laundry, storage area, and bedroom. The areas of the individual rooms are 298, 92, 81, and 150 square feet respectively. What is the total area of these rooms?  _____
   (1)

2. A contractor has $3,575 in a checking account and writes checks for $72 and $139. What is the balance remaining in the account?  _____
   (2)

3. Find the total length of 15 pieces of chair rail if each piece is 65 inches long.  _____
   (3)

4. At a rate of 82 square feet per hour, how long will it take to lay 738 square feet of subfloor?  _____
   (4)

5. Determine the total length of the piece of stock illustrated.
   (10)

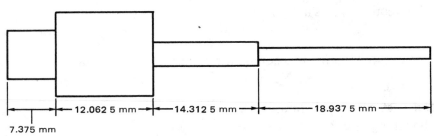

|←—— 12.062 5 mm ——→|←— 14.312 5 mm —→|←——— 18.937 5 mm ———→|

7.375 mm

   _____

6. Determine the final thickness of a 2 1/4-inch piece of stock if 3/16 inch is taken off both surfaces.  _____
   (6)

7. It takes 1/4 hour to place 9 linear feet of sill. How long does it take to place 126 feet?  _____
   (25)

8. How many risers, each 8 3/8 inches, are there in a flight of stairs with a total rise of 5'-7''?  _____
   (8)

9. A table top 0.750 inch thick is covered with laminated plastic 0.0625 inch thick. What is the total thickness of the top?
(10)

_____

10. Find the length of the bottom rail of the illustrated table.
(11)

11. It is estimated that a house measuring 26'-0" x 48'-0" can be built for $33.50 per square foot. Find the total estimated cost of this house.
(12, 19)

_____

12. The labor cost of laying 18 squares of fiberglass shingles is $175.50. What is the cost of laying one square?
(13, 25)

_____

13. Express the fraction 7/8 as a decimal fraction.
(14)

_____

14. Express the decimal 0.6875 as a common fraction.
(14)

_____

15. The actual cost of a small house is $14,375. At a profit rate of 8%, what is the selling price?
(15)

_____

16. A contractor borrowed $30,000 at a yearly interest rate of 12%. What does this loan cost per year?
(16)

_____

17. If the list price of a framing square shown in a catalog is $11.99, subject to a 25% discount, what must a carpenter pay for this square?
(17)

_____

18. Find the number of feet of crown molding to be ordered for two rooms measuring 11'-8" x 12'-6" and 14'-0" x 16'-6".
(18)

_____

19. Determine the area (in square feet) of the vented portion of the louver pictured.
(19, 29)

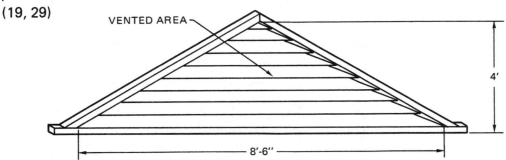

20. How many board feet of material are there in a piece of stock measuring 2" x 8" x 16'?
(20)

_____

21. Determine the number of cubic feet taken up by a stairwell measuring 9'-6" x 3'-4" x 8'-6".
(21, 32)

_____

22. If a 10" x 4 3/4" I beam weighs 35 pounds per linear foot, what is the weight of a beam 24 feet long?
(3, 25)

_____

23. Determine the diameter of the small pulley in the illustration.
(25)

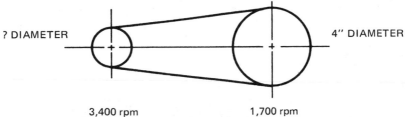

_____

24. Express each of the following ratios in simplest form:
(23)

a. 90:360

_____

b. 36:24

_____

c. 150:25

_____

25. Find the number of cubic feet of concrete contained in a footing which is 2 feet square and 1 foot thick.
(21, 32)

_____

26. What stringer length is required for a stairway with a total rise of 12'-0" and a total run of 16'-0"?
    (27, 47)

    _____

27. Determine the cross-sectional area of illustrated objects **A** and **B**.
    (19, 28)

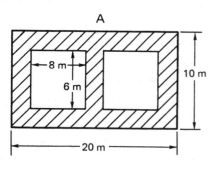

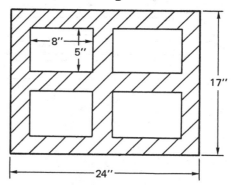

    **A** _____

    **B** _____

28. Find the area of the gable end of a house which has a span of 33'-0" and a 1/3 pitch.
    (23, 29)

    _____

29. Determine, to the nearest hundredth, the area of a circle with a diameter of 5.25 cm.
    (19, 31)

    _____

30. From the illustration, determine the cubic footage of the storage area shown. The walls are 7'-6" high.
    (21, 32)

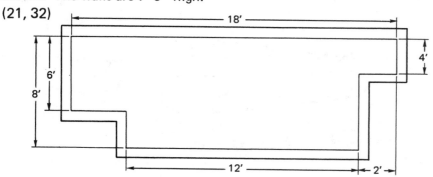

    _____

31. What is the capacity, to the nearest hundredth cubic metre, of a circular storage tank with a diameter of 6 m and a height of 14.1 m?
    (21, 33)

    _____

32. Find the number of linear feet of sill plate needed for a house measuring 28'-6" x 54'-0".
    (18, 35)

    _____

33. Determine the cost of 3 pieces of 1" x 8" x 8' oak at a price of $1.15 per board foot.
    (20)

    _____

# ACHIEVEMENT REVIEW B

**Note:** The numbers in parentheses, ( ), given below each question show the unit or units in which similar problems have been discussed.

1.  A contractor paid material bills of $2,760, $1,128, $765, and $4,385. What is the total cost of these materials?

    (1)

    _____

2.  A piece of stock 47 inches long is cut from a board 72 inches long. What is the length of the remaining piece? Do not allow for saw kerf.

    (2)

    _____

3.  If a carpenter can place 63 linear feet of joists per hour, how many feet are placed in an 8-hour day?

    (3, 25)

    _____

4.  Determine the number of rafters that can be cut from a piece of stock 216 inches long if each rafter is 36 inches long. Do not allow for saw kerf.

    (4)

    _____

5.  Find the total length of the wall section shown.

    (5)

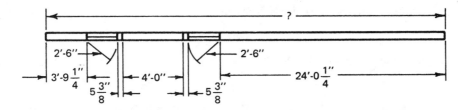

    _____

6.  After being covered with 1/16-inch laminated plastic, a counter top measures 13/16 inch thick. What is the thickness of the core stock?

    (6)

    _____

7.  In a flight of stairs containing 14 risers, each 7 1/8 inches, what is the total rise of the stairway?

    (7)

    _____

8.  One carpenter can lay 12 1/2 squares of wood shingles in 4 1/2 days. How many squares are laid in an average day?

    (8, 25)

    _____

9. An expense record for one week is as follows:   material $4,328.50, labor $878.65, overhead $208.63.  What are the total expenses?
   (10)                                                                          _____

10. What is the length of distance **A** of the pattern illustrated?
    (11, 13)

11. At a rate of $45 per square foot, find the estimated cost of a house which measures 32'-0" x 56'-0".
    (12, 19, 28)                                                                 _____

12. If the cost of 4500 board feet of lumber is $1,102.50, what is the cost of 1000 board feet?
    (13, 25)                                                                     _____

13. Express the common fraction 3/8 as a decimal fraction.
    (14)                                                                         _____

14. Express the decimal 0.775 as a common fraction.
    (14)                                                                         _____

15. If the actual cost of a certain job is $6,600, what is the selling price at a profit rate of 12%?
    (15)                                                                         _____

16. If a contractor borrows $4,250.00 at an interest rate of 13% per year, what is the yearly interest cost?
    (16)                                                                         _____

17. A materials bill is $1,926.90, less 2% if paid in 30 days.  What is the actual payment if the bill is paid one week after ordering?
    (17)                                                                         _____

18. Determine the number of linear feet of base shoe needed for two rooms, each measuring 12'-6" x 14'-0".  (Make no allowance for openings.)
    (18, 40)                                                                     _____

19. What is the area in square feet, of the vented portion of the louver pictured?
    (19, 29)

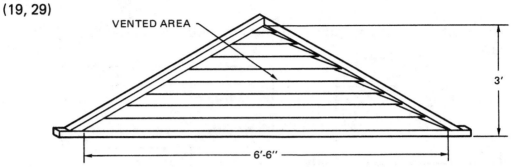

VENTED AREA

3'

6'-6''

_____

20. How many board feet are there in a piece of stock 3/4'' x 8'' x 48''?     _____
    (20)

21. How many cubic metres of storage space are there in an area measuring 7 m x 16 m x 8 m?     _____
    (21, 32)

22. It is estimated that 20 pounds of 8d common nails are required for 1000 board feet of subfloor. Determine the number of pounds needed for 16,000 board feet.     _____
    (22, 25)

23. Express each of the following ratios in simplest form:
    (23)

    a. 15:25     _____

    b. 7:42     _____

    c. 18:36     _____

24. What is the speed in revolutions per minute of the small pulley illustrated?
    (25)

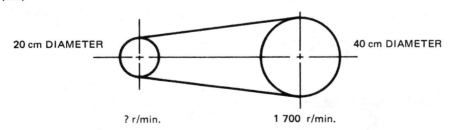

20 cm DIAMETER          40 cm DIAMETER

? r/min.          1 700 r/min.

_____

25. Find the volume of a storage tank which is 4 feet square and 4 feet high.
    (21, 32)     _____

26. What stringer length is needed for a flight of stairs with a total rise of 7'-6'' and a total run of 5'-2''?
(27)

_____

27. From the illustration, find the cross-sectional area of objects A and B.
(19, 28)

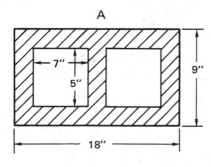

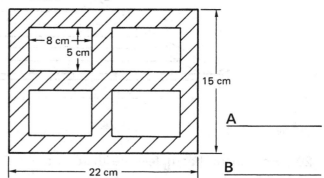

A _____

B _____

28. Calculate the area of the gable end of a house which has a span of 30'-6'' and a rise of 7'-6''.
(19, 29)

_____

29. Determine the area of a circle with a diameter of 6 inches.
(19, 31)

_____

30. What is the volume of the illustrated step block?
(21, 32)

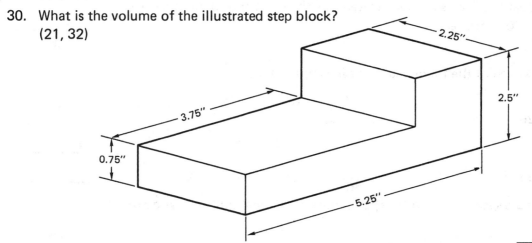

_____

31. Find the volume of a circular cistern with a diameter of 0.6 m and height of 1.3 m.
(21, 33)

_____

32. How many linear feet of sill plate are required for a house which measures 34'-6'' x 62'-0''?
(18, 35)

_____

33. What is the cost of 4 pieces of pine 1'' x 9'' x 8' at a price of 98¢ a board foot?
(20)

_____

# DIAGNOSTIC READING SURVEY

**To The Teacher:**   Detailed instructions for administering this survey and interpreting the results are provided in the Instructor's Guide. Please follow them exactly.

**To The Student:**   This survey is to see what basic skills you already have and what skills you may need to improve in this program area. The information gained from this survey will help your instructor plan the kind of instruction that is best for you.

This is a timed survey. If you complete a section before time is called, you may not go on to the next section and you may not return to a previous section. You may go over your answers in the section you have just completed. When you are sure of your answers, turn your answer sheet over. If everyone completes a section before the time limit, the timing may be stopped.

The first section is Vocabulary. This section measures how well you know the meaning of words. It contains ten items and you will have ten minutes to complete your work.

The second section is Spelling. This section contains ten items and you will have five minutes to complete your work.

The third section is Comprehension and will measure your knowledge of reference information and how well you understand what you read. It contains twenty-five items and you will have forty-five minutes to complete your work.

The complete survey takes one hour. Follow all directions carefully. Write answers neatly in the correct spaces on your answer sheet.

This reading survey has been especially prepared by Carla Hoke for students interested in carpentry.

The Competency-based Reading Survey is designed to determine skill levels in the following areas: Vocabulary, Spelling, and Comprehension. The subskills in the Comprehension section include Reference Skills, Using a Table of Contents and an Index, Finding the Main Idea, Locating Facts, Following Directions, and Reading Charts and Diagrams.

Item analysis sheets will provide specific information on each student's reading strengths and weaknesses and will aid both the reading specialist and the vocational instructor in planning developmental and corrective programs and in individualizing instruction.

Each student will need:
- the Carpentry Reading Survey
- an answer sheet
- a pencil or pen

## VOCABULARY

**DIRECTIONS:**   Look at the sample test word below. The word is *beam*. Now read the definitions just below *beam*. Find the one definition in this group that means most nearly the same as *beam*. Definition "b" is the correct definition, "structure used between posts, columns, and walls". On your answer sheet, under the section marked Vocabulary, find the line marked "Sample" and write a "b" on that line.

For each numbered word in this section, write the letter of the definition that means most nearly the same as the given word on your answer sheet. If you are not sure of an answer, make the most careful guess you can.

— — — — — — — — — — — — — — — — — — — — — — —

**Sample:**   beam
  a. material used to make concrete
  b. structure used between posts, columns, and walls
  c. type of tile
  d. type of door hinge
  e. none of the above

STOP

— — — — — — — — — — — — — — — — — — — — — — —

Do not turn the page until you are told to begin.

1.  truss
    a. horizontal part of a step
    b. trim required to finish one side of a door
    c. structural unit which provides support over wide spans
    d. a small porch
    e. none of the above

2.  partition
    a. term used to indicate nail length
    b. slope of a roof
    c. the underside of an eave or cornice
    d. wall that subdivides space
    e. none of the above

3.  plumb
    a. perpendicular to the floor
    b. parallel to the floor
    c. horizontal
    d. installed fixtures of a building
    e. none of the above

4.  joist
    a. top and two sides of a door or window frame
    b. one of a series of parallel framing members used to support floor and ceiling loads
    c. extension of a sill
    d. metal trough
    e. none of the above

5.  alteration
    a. change
    b. an interior molding
    c. a railing
    d. a filling of mortar
    e. none of the above

6.  bevel
    a. an offer to supply, at a specified price
    b. a defect
    c. a type of stain
    d. to cut to an angle other than a right angle
    e. none of the above

7.  butt
    a. replacement of earth
    b. type of roof
    c. type of door hinge
    d. a temporary framework
    e. none of the above

8. casing
    a. trimming around a door or window
    b. wood seasoned by exposure to sun and air
    c. a residue from evaporated petroleum
    d. a protecting support for a shelf
    e. none of the above

9. knot
    a. branch or limb embedded in the tree
    b. roof which rises from all 4 sides of a building
    c. sheet metal used in roof construction
    d. a low wall along the edge of a roof
    e. none of the above

10. column
    a. vertical stair part
    b. upright supporting member, circular or rectangular in shape
    c. a temporary structure to support workmen
    d. a plan of a city indicating boundaries
    e. none of the above

**STOP**

— — — — — — — — — — — — — — — — — — — — — —

Turn your answer sheet over and wait for further directions.


## SPELLING

**DIRECTIONS:** Look at the sample below. Only one word is correctly spelled. The correct spelling is "d", level. On your answer sheet, under the section marked Spelling, find the line marked "Sample" and write a "d" on that line.

For each numbered problem in this section, write the letter of the word which is correctly spelled on your answer sheet. If you are not sure of an answer, make the most careful guess you can.

— — — — — — — — — — — — — — — — — — — — — —

**Sample:**
    a. levil
    b. leval
    c. levle
    d. level

**STOP**

— — — — — — — — — — — — — — — — — — — — — —

Do not turn the page until you are told to begin.

1.  a. foundasion
    b. foundation
    c. foundashun
    d. foundayshun

2.  a. seeling
    b. ceeling
    c. ceiling
    d. cieling

3.  a. gutter
    b. gudder
    c. guddar
    d. guttar

4.  a. hamer
    b. hammar
    c. hamar
    d. hammer

5.  a. chisle
    b. chisel
    c. chissle
    d. chizzle

6.  a. construcktion
    b. construction
    c. construksion
    d. construcshun

7.  a. shingle
    b. shingel
    c. shingal
    d. shungil

8.  a. panal
    b. panle
    c. panile
    d. panel

9.  a. gerage
    b. geraje
    c. garage
    d. garadge

10. a. plywood
    b. pliwood
    c. plywould
    d. pleyewood

**STOP**

— — — — — — — — — — — — — — — — — — —

**PRACTICE SPACE**

# COMPREHENSION

**DIRECTIONS:** This section of the test consists of two parts. The items in the first part measure your knowledge of reference information. The items in the second part show how well you understand what you read.

This section contains 25 items. Choose the best answer for the items in the first part. In the second part, read the information given and do the items that follow. Fill in the space on your answer sheet with the letter of the answer you choose.

— — — — — — — — — — — — — — — — — — — — —

**Sample:** Read the passage below.

Joe decided to get a part-time job after school. He looked in the classified section of the newspaper since this part contains the help wanted notices. He found two companies near home which were hiring and called for an appointment for an interview.

Now read the item below.

The classified section of the newspaper contains

a. secret information
b. help wanted notices
c. current news stories
d. advice to the lovelorn

Look at the section of your answer sheet under the word Comprehension. Put a "b" on the line for the answer for the Sample item because the classified section of the newspaper contains "help wanted notices".

**STOP**

— — — — — — — — — — — — — — — — — — — — —

Do not turn the page until you are told to begin.

Items 1–3 measure your knowledge of reference information. Read each question and mark your answer on the answer sheet.

1. A glossary contains
   a. definitions
   b. quotes from famous people
   c. tables and diagrams
   d. an overview of the book

2. The appendix of a book consists of
   a. unusual words used in the text
   b. a list of chapter headings
   c. a list of authors
   d. supplementary material

3. A bibliography contains
   a. a list of important names and dates
   b. a list of books
   c. a list of topics covered in the text
   d. a list of words used in the text

Now read the Table of Contents. Answer the questions which follow.

| TABLE OF CONTENTS | |
|---|---|
| CHAPTER | PAGE |
| 1. Hand Tools | 1 |
| 2. Power Tools | 27 |
| 3. Building Materials | 53 |
| 4. Plans, Specifications, Codes | 71 |
| 5. Types of Insulation | 87 |
| 6. Roofing Materials | 135 |
| 7. Exterior Trim | 185 |
| 8. Interior Trim | 277 |
| 9. Carpentry as a Career | 451 |

4. "Types of Insulation" begins on page
   a. 53
   b. 71
   c. 78
   d. 87

5. Page 177 is in which chapter?
   a. Plans, Specifications, Codes
   b. Roofing Materials
   c. Interior Trim
   d. Types of Insulation

6. The most complete information on reading blueprints can be found in chapter
   a. 3
   b. 4
   c. 8
   d. 9

Read the Index on the next page. Then do items 7–9 and mark your answers on the answer sheet.

```
┌─────────────────────────────────────────────┐
│                   INDEX                      │
│                                              │
│  Cabinet                                     │
│          basic framing, 386                  │
│          built-ins, 400, 401                 │
│          construction procedures, 386        │
│          dimensions, factory-built, 404      │
│          doors, 394                          │
│          facing, 388                         │
│          hardware, 399                       │
│  Carpentry                                   │
│          apprenticeship, 452                 │
│          as a career, 450                    │
│          opportunities, 451                  │
│          personal qualifications, 453        │
│  Circular saws, 27                           │
│  Clamping tools, 19                          │
│                                              │
└─────────────────────────────────────────────┘
```

Refer to this Index to answer questions 7–9.

7. If you want to know how to build cabinets, you can look on page
   a. 386
   b. 400
   c. 399
   d. 404

8. Job possibilities in carpentry are discussed on page
   a. 452
   b. 451
   c. 453
   d. 415

9. Factory-built cabinets are discussed on page
   a. 401
   b. 388
   c. 400
   d. 404

Read the passages which begin on the next page. After each passage, choose the best answer for each item and mark your answer on the answer sheet.

Asphalt roofing shingles are made from felt that is saturated and coated with asphalt. The felt is made from products containing cellulose fibers (rags, paper, and wood). Asphalt is a by-product of the petroleum industry and goes by the trade name of Asphalt Flux. The saturation of the asphalt increases the preserving and waterproofing properties of the asphalt shingle.

In addition to the asphalt, flux mineral stabilizers are sometimes used to coat asphalt shingles. It has been found that the products will better resist effects of the weather and generally be more stable if they contain a small percentage of minerals such as silica, slate dust, talc, dolomite, and trap rock. The mineral granules also reflect the rays of the sun, increase the fire resistance of the shingle, and enhance the architectural style of the home, primarily because of the variety of colors and color blends.

Asphalt shingles are manufactured in weights from 145 pounds to 330 pounds per square. A square is an area which measures 10 feet by 10 feet, or 100 square feet. Three or four bundles of 36-inch shingles will cover a square. The most common weights of the asphalt composition shingle used in light construction are 235 and 240 pounds. As shingles increase in weight, the life expectancy of the shingle also increases.

Answer questions 10–13.

10. The best title for this selection is
    a. Applying Asphalt Shingles        c. About Asphalt Shingles
    b. Treatment of Asphalt Shingles     d. Maintenance of Asphalt Shingles

11. "Dolomite" is
    a. a trade name for a type of asphalt shingle
    b. a measure of weight
    c. a by-product of asphalt
    d. a mineral

12. Mineral granules in asphalt
    a. absorb the sun's rays
    b. enhance the architectural style of the home
    c. decrease the fire resistance of the shingle
    d. all of the above

13. The heaviest weight of asphalt shingles is
    a. 330 lbs. per square yard        c. 330 lbs. per square
    b. 330 lbs. per square foot         d. 330 lbs. per square inch

Read the passage on the next page and answer questions 14–16. Put your answers on the answer sheet.

## HOW TO INSTALL SOLID WOOD PANELING

Note:   The wall must be prepared to receive paneling before it is installed.  If the paneling is installed vertically, furring or blocking must be used.  Two rows of blocking should be provided if it is used.  If it is necessary to use furring strips, a strip is placed at the top and bottom of the wall with strips spaced between them.

1. If the wall is not plumb, shim the furring strips sufficiently.
2. Starting at one corner, place the first panel in position.  Check to be sure it is plumb.
3. Secure the position of the panel by face nailing with 6d finishing nails.  The rest of the nails are blind nailed, eliminating the need for countersinking and filling.  Only the first and last panels are face nailed.
4. Place the second board in position and blind nail it.
5. Install the remainder of the panels in the same manner.  Note:  Avoid ending with a molded edge in a corner.
6. Interior corners are butted together to form a tight joint.  Exterior corners may be either fitted with corner moldings or mitered to form a finished corner.  Note:  The joint between the paneling and the ceiling and the joint between the paneling and the floor are usually covered by molding.

Answer questions 14–16.

14. When installing paneling, most of the nails are
    a. countersunk
    b. blind nailed
    c. filled
    d. face nailed

15. At each corner,
    a. butt together panels to form a tight joint.
    b. fit with corner moldings.
    c. avoid ending with a molded edge.
    d. miter to form a finished corner.

16. According to the passage, to prepare a wall to receive paneling,
    a. install furring or blocking.
    b. clean the wall with detergent.
    c. make sure you have purchased enough material.
    d. read the directions carefully.

Read the passage on the next page and answer questions 17-20.  Put your answers on the answer sheet.

Paneling can provide an increase in the structural strength of a room besides being decorative. The three basic types of paneling are hardboard, plywood, and solid wood.

Wood fibers are subjected to heat and pressure and then shaped to form hardboard paneling. The paneling is smoother, harder, and denser than most other types of paneling and is highly resistant to moisture and dents. The paneling is usually 1/8 or 1/4 inch thick and is available in various sizes.

The panels have varying surface finishes to complement the decor of any room. They are available in simulated wood-grain patterns of walnut, pecan, chestnut, teak, oak, cherry, and birch. Other panels have surfaces that are marbleized, burlapped, louvered, embossed and covered with lacy prints. The panels may also be purchased in a plain design and painted any color. If a particular building code requires a flame-retardant surface, there are hardboard panels that are *impregnated* with fire-retardant chemicals.

Most hand and power tools will easily cut hardboard paneling. Some carpenters prefer to keep a special blade for cutting hardboard, since hardboard is tougher than most natural woods and has a tendency to destroy the *set* of a blade. The set is the angle of the saw teeth. A crosscut saw is used for straight cuts, and a coping saw is used for curved cuts.

Conventional wood fasteners or adhesives may be used to secure the paneling to framing members. If plain finishing nails are used they should be set and the holes filled with colored putty. Color-coordinated nails may be purchased to match paneling.

Answer questions 17-20.

17. The best title for this selection is
    a. Types of Paneling
    b. Advantages of Paneling
    c. Installation of Paneling
    d. Hardboard Paneling

18. According to the selection, an advantage of hardboard paneling is that
    a. it has a rustic appearance
    b. it resists moisture
    c. it contains V-grooves to help conceal nails
    d. it absorbs heat

The words "set" and "impregnated" were used in the passage you just read. Use the back of your answer sheet to write a definition of each word.

19. As used in the passage, "set" means _____.

20. As used in the passage, "impregnated" means _____.

On the next page you will find a perspective drawing. Use that drawing to answer questions 21-23. Put your answers on the answer sheet.

## PERSPECTIVE DRAWING

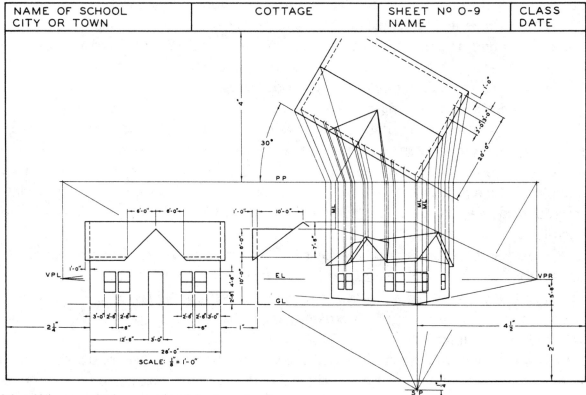

21. What scale is used in this drawing?
    a. 1'-0'' = 1/8''
    b. 1/8'' = 1'-0''
    c. 1/4'' = 1'-0''
    d. 4'' = 1'-0''

22. How far above the foundation is the top of each window?
    a. 4'-6''
    b. 2'-6''
    c. 7'
    d. 3'-0''

23. How wide is the overhang of the roof?
    a. 6'-0''
    b. 1'-0''
    c. 3'-0''
    d. 8'-0''

On the next page you will find a chart which gives powers and roots of numbers. Use this chart to answer questions 24-25. Put your answers on the answer sheet.

## POWERS AND ROOTS OF NUMBERS (1 through 100)

| Num-ber | Powers | | Roots | | Num-ber | Powers | | Roots | |
|---|---|---|---|---|---|---|---|---|---|
| | Square | Cube | Square | Cube | | Square | Cube | Square | Cube |
| 1 | 1 | 1 | 1.000 | 1.000 | 51 | 2,601 | 132,651 | 7.141 | 3.708 |
| 2 | 4 | 8 | 1.414 | 1.260 | 52 | 2,704 | 140,608 | 7.211 | 3.733 |
| 3 | 9 | 27 | 1.732 | 1.442 | 53 | 2,809 | 148,877 | 7.280 | 3.756 |
| 4 | 16 | 64 | 2.000 | 1.587 | 54 | 2,916 | 157,464 | 7.348 | 3.780 |
| 5 | 25 | 125 | 2.236 | 1.710 | 55 | 3,025 | 166,375 | 7.416 | 3.803 |
| 6 | 36 | 216 | 2.449 | 1.817 | 56 | 3,136 | 175,616 | 7.483 | 3.826 |
| 7 | 49 | 343 | 2.646 | 1.913 | 57 | 3,249 | 185,193 | 7.550 | 3.849 |
| 8 | 64 | 512 | 2.828 | 2.000 | 58 | 3,364 | 195,112 | 7.616 | 3.871 |
| 9 | 81 | 729 | 3.000 | 2.080 | 59 | 3,481 | 205,379 | 7.681 | 3.893 |
| 10 | 100 | 1,000 | 3.162 | 2.154 | 60 | 3,600 | 216,000 | 7.746 | 3.915 |
| 11 | 121 | 1,331 | 3.317 | 2.224 | 61 | 3,721 | 226,981 | 7.810 | 3.936 |
| 12 | 144 | 1,728 | 3.464 | 2.289 | 62 | 3,844 | 238,328 | 7.874 | 3.958 |
| 13 | 169 | 2,197 | 3.606 | 2.351 | 63 | 3,969 | 250,047 | 7.937 | 3.979 |
| 14 | 196 | 2,744 | 3.742 | 2.410 | 64 | 4,096 | 262,144 | 8.000 | 4.000 |
| 15 | 225 | 3,375 | 3.873 | 2.466 | 65 | 4,225 | 274,625 | 8.062 | 4.021 |
| 16 | 256 | 4,096 | 4.000 | 2.520 | 66 | 4,356 | 287,496 | 8.124 | 4.041 |
| 17 | 289 | 4,913 | 4.123 | 2.571 | 67 | 4,489 | 300,763 | 8.185 | 4.062 |
| 18 | 324 | 5,832 | 4.243 | 2.621 | 68 | 4,624 | 314,432 | 8.246 | 4.082 |
| 19 | 361 | 6,859 | 4.359 | 2.668 | 69 | 4,761 | 328,509 | 8.307 | 4.102 |
| 20 | 400 | 8,000 | 4.472 | 2.714 | 70 | 4,900 | 343,000 | 8.367 | 4.121 |
| 21 | 441 | 9,261 | 4.583 | 2.759 | 71 | 5,041 | 357,911 | 8.426 | 4.141 |
| 22 | 484 | 10,648 | 4.690 | 2.802 | 72 | 5,184 | 373,248 | 8.485 | 4.160 |
| 23 | 529 | 12,167 | 4.796 | 2.844 | 73 | 5,329 | 389,017 | 8.544 | 4.179 |
| 24 | 576 | 13,824 | 4.899 | 2.884 | 74 | 5,476 | 405,224 | 8.602 | 4.198 |
| 25 | 625 | 15,625 | 5.000 | 2.924 | 75 | 5,625 | 421,875 | 8.660 | 4.217 |
| 26 | 676 | 17,576 | 5.099 | 2.962 | 76 | 5,776 | 438,976 | 8.718 | 4.236 |
| 27 | 729 | 19,683 | 5.196 | 3.000 | 77 | 5,929 | 456,533 | 8.775 | 4.254 |
| 28 | 784 | 21,952 | 5.292 | 3.037 | 78 | 6,084 | 474,552 | 8.832 | 4.273 |
| 29 | 841 | 24,389 | 5.385 | 3.072 | 79 | 6,241 | 493,039 | 8.888 | 4.291 |
| 30 | 900 | 27,000 | 5.477 | 3.107 | 80 | 6,400 | 512,000 | 8.944 | 4.309 |
| 31 | 961 | 29,791 | 5.568 | 3.141 | 81 | 6,561 | 531,441 | 9.000 | 4.327 |
| 32 | 1,024 | 32,798 | 5.657 | 3.175 | 82 | 6,724 | 551,368 | 9.055 | 4.344 |
| 33 | 1,089 | 35,937 | 5.745 | 3.208 | 83 | 6,889 | 571,787 | 9.110 | 4.362 |
| 34 | 1,156 | 39,304 | 5.831 | 3.240 | 84 | 7,056 | 592,704 | 9.165 | 4.380 |
| 35 | 1,225 | 42,875 | 5.916 | 3.271 | 85 | 7,225 | 614,125 | 9.220 | 4.397 |
| 36 | 1,296 | 46,656 | 6.000 | 3.302 | 86 | 7,396 | 636,056 | 9.274 | 4.414 |
| 37 | 1,369 | 50,653 | 6.083 | 3.332 | 87 | 7,569 | 658,503 | 9.327 | 4.481 |
| 38 | 1,444 | 54,872 | 6.164 | 3.362 | 88 | 7,744 | 681,472 | 9.381 | 4.448 |
| 39 | 1,521 | 59,319 | 6.245 | 3.391 | 89 | 7,921 | 704,969 | 9.434 | 4.465 |
| 40 | 1,600 | 64,000 | 6.325 | 3.420 | 90 | 8,100 | 729,000 | 9.487 | 4.481 |
| 41 | 1,681 | 68,921 | 6.403 | 3.448 | 91 | 8,281 | 753,571 | 9.539 | 4.498 |
| 42 | 1,764 | 74,088 | 6.481 | 3.476 | 92 | 8,464 | 778,688 | 9.592 | 4.514 |
| 43 | 1,849 | 79,507 | 6.557 | 3.503 | 93 | 8,649 | 804,357 | 9.644 | 4.531 |
| 44 | 1,936 | 85,184 | 6.633 | 3.530 | 94 | 8,836 | 830,584 | 9.695 | 4.547 |
| 45 | 2,025 | 91,125 | 6.708 | 3.557 | 95 | 9,025 | 857,375 | 9.747 | 4.563 |
| 46 | 2,116 | 97,336 | 6.782 | 3.583 | 96 | 9,216 | 884,736 | 9.798 | 4.579 |
| 47 | 2,209 | 103,823 | 6.856 | 3.609 | 97 | 9,409 | 912,673 | 9.849 | 4.595 |
| 48 | 2,304 | 110,592 | 6.928 | 3.634 | 98 | 9,604 | 941,192 | 9.900 | 4.610 |
| 49 | 2,401 | 117,649 | 7.000 | 3.659 | 99 | 9,801 | 970,299 | 9.950 | 4.626 |
| 50 | 2,500 | 125,000 | 7.071 | 3.684 | 100 | 10,000 | 1,000,000 | 10.000 | 4.642 |

24. The square root of 78 is
   a. 6,084
   b. 474,552
   c. 8.832
   d. 4.273

25. The number whose square root is 4.899 is
   a. 2.884
   b. 13,824
   c. 576
   d. 24

**STOP**

DIAGNOSTIC READING SURVEY
ANSWER SHEET

NAME _____

PROGRAM _____

DATE _____

| VOCABULARY | SPELLING | COMPREHENSION |
|---|---|---|
| Sample: _____ | Sample: _____ | Sample: _____ |
| 1. _____ | 1. _____ | 1. _____ |
| 2. _____ | 2. _____ | 2. _____ |
| 3. _____ | 3. _____ | 3. _____ |
| 4. _____ | 4. _____ | 4. _____ |
| 5. _____ | 5. _____ | 5. _____ |
| 6. _____ | 6. _____ | 6. _____ |
| 7. _____ | 7. _____ | 7. _____ |
| 8. _____ | 8. _____ | 8. _____ |
| 9. _____ | 9. _____ | 9. _____ |
| 10. _____ | 10. _____ | 10. _____ |
|  |  | 11. _____ |
|  |  | 12. _____ |
|  |  | 13. _____ |
|  |  | 14. _____ |
|  |  | 15. _____ |
|  |  | 16. _____ |
|  |  | 17. _____ |
|  |  | 18. _____ |
|  |  | 19. _____ |
|  |  | 20. _____ |
|  |  | 21. _____ |
|  |  | 22. _____ |
|  |  | 23. _____ |
|  |  | 24. _____ |
|  |  | 25. _____ |

Do not write in this space!

SCORES:

Vocabulary: _____

Spelling: _____

Comprehension: _____

MAY HAVE DIFFICULTY WITH:

Vocabulary _____

Spelling _____

Reference Skills _____

Table of Contents _____

Using an Index _____

Finding Main Idea _____

Locating Facts _____

Vocabulary in Context _____

Following Directions _____

Reading Charts _____

Reading Tables _____

# GLOSSARY

**Asbestos Shingles** — Fireproof shingles made from asbestos and portland cement.

**Asphalt** — A black mineral pitch used for waterproofing roads, driveways, walks, and to impregnate felt when used as roofing and building paper.

**Asphalt Shingles** — Shingles made from asphalt-saturated felt and coated on the surface with various colored minerals. They are manufactured both in strips and individual shingles.

**Backfill** — Dirt placed in the trench around a foundation.

**Baffles** — Plates used to regulate the flow of materials between adjacent compartments.

**Baseboard** — Finishing board used at the base of a vertical wall.

**Batter Board** — The temporary frame used to locate foundation corners.

**Beam** — An inclusive term for joists, girders, and rafters.

**Bed Molding** — Molding used to cover the joint between the plancier and frieze.

**Bevel Siding** — Wedge shaped boards used as finish siding on the exterior of a structure.

**Board Foot** — The equivalent of a board 1 inch thick, 1 foot wide, and 1 foot long. It equals 144 cubic inches.

**Bridging** — Diagonal bracing fitted between joists to reinforce a floor. It is generally installed in crossing pairs and often called bridging.

**Casement Sash** — Windows which are hinged on one side.

**Casing** — The trim around a door or window, either outside or inside. The finish lumber covering a rough-sawed post is also called casing.

**Cement** — A fine grayish powder produced by burning in a kiln a carefully blended mixture of limestone, clay, or shale. The clinkers thus formed are pulverized and bagged.

**Chair Rail** — Horizontal moldings placed on a wall to protect the wall from damage by furniture.

**Circumference** — The distance around the outside of a circle.

**Cistern** — A large tank for holding water.

**Clapboards** — Boards used as exterior siding, placed horizontally and lapped.

**Clothes Chute** — A chute designed for the purpose of dropping soiled clothes from an upper level to a basement laundry.

**Collar Beam** — A beam, above the wall plate, used to tie two rafters together.

**Common Rafter** — Main structural roof support running at right angles from the wall plate to the ridge.

**Concrete** — A building material made from a mixture of cement, sand, gravel, and water.

**Corner Mold** — Molding applied to exterior corners of a room to protect the plaster or dry wall board.

**Cornice** — The section of exterior trim from the top of the wall to the projection of the rafters.

**Crown Molding** — A molding used to conceal the joint formed by the ceiling and wall. It is placed in contact with both ceiling and wall.

**Culvert** — Arched channel for the passage of water below ground level.

**Dado** — A rectangular cut in a piece of wood across the wood grain.

**Dead Load** — The weight of all stationary construction included in a building.

**Diameter** — A line passing through the center of a circle and terminated at both ends by the circumference. It is equal in distance to twice the radius.

**Dry Wall** — A wall covering of stone or other durable material, laid without mortar. Gypsum board is a common type of dry wall.

**Fascia** — A board nailed to the ends of the rafters.

**Fiberboard** — Sheet material formed from cellulose or wood fibers pressed together.

**Fire Stop** — A block or closure, between the studs or joists, to prevent the spread of fire or smoke, also called a draft stop.

**Flashing** — Material used around chimneys, vents, windows, etc. to prevent moisture from entering the building. Sheet metal is a very common type of flashing.

**Flush** — Two members forming an even surface.

**Footing** — Enlargement at the base of a pier, column, or foundation wall.

**Form** — A wood or metal structure constructed for the purpose of retaining fresh concrete until it sets.

**Foundation** — The supporting portion of a structure usually below grade. It is composed of two parts, footing and foundation wall.

**Framing Square** — A right angle tool used in framing buildings. It is generally made of steel and includes tables in addition to the scales.

**Framing Timber** — Lumber used in framing with dimensions greater than 4 inches by 6 inches.

**Frieze** — Decorative trim on the upper part of a wall connecting the top of the siding with the soffit.

**Gable** — The portion of a wall bound by the plate and the inclined roof of a building.

**Gable Roof** — A roof formed by common rafters. It slopes up from only two walls.

**Gambrel Roof** — A gable roof each slope of which is broken into two planes with different pitches.

**Girder** — A large horizontal beam used to support smaller beams or joists.

**Glazing** — The process of installing glass into a sash frame.

**Glazing Point** — Small piece of metal used to secure glass in a window prior to puttying.

**Grout** — A watery mixture of sand and cement.

**Hardboard** — Pressed wood panels made from wood fibers, not wood chips. It is normally bonded by the natural materials in wood.

**Header** — IN FLOOR FRAMING: A short traverse or horizontal joist which supports the ends of one of more cut off joists. IN WALL FRAMING: Supporting member for the load over the opening for a window or door, also called lintel.

**Hip Rafter** — A rafter at the intersection of two roof slopes forming an external angle or a hip. It extends diagonally from the corner of the plate to the ridge.

**Hip Roof** — A roof which slopes up toward the center or ridge from four sides.

**Hypotenuse** — The longest side of a right triangle. It is the side opposite the 90 degree angle.

**Isosceles Triangle** — A triangle with two equal sides.

**Jamb** — The vertical posts of door or window framing, also frequently used to refer to the vertical casing within a doorway or window.

**Joint** — The place where two or more members come together.

**Joist** — Beam used to support a floor or ceiling.

**Lally Column** — A steel post used to support the main girders in a building.

**Laminated Plastic** — Layered plastic bound with resin, frequently used as a surface on counter tops.

**Lath** — Wood strips, metal mesh or gypsum used as a foundation for plaster.

**Leg of Triangle** — The sides of a right triangle which form the right angle.

**Linear Foot** — A measure of twelve inches along a straight line.

**Line Lengths** — The hypotenuse of a triangle whose base is the total run and whose altitude is the total rise.

**Load Carrying Partitions** — Partitions which support the floor joists above them.

**Live Load** — Movable load such as people, furnishings, etc. as contrasted to dead load which is a permanent part of a building.

**Louver** — Ventilation opening.

**Masonry** — Materials bonded together with mortar to form a wall.

**Masonry Cement** — A mixture of Portland cement and lime.

**Matched Boards** — Boards that are edge dressed and shaped into tongue-and-groove edges to join together snugly.

**Miter Box** — A device used to guide a saw when cutting at an angle.

**Mortise** — A hole or slot into which some other part or tenon fits.

**Newel Post** — A vertical post at the foot of a stairway into which the handrail is fitted.

**Novelty Siding** — Siding which has been processed to give an unusual design. This siding can serve as both siding and sheathing.

**On Center** — A term which refers to the distance from the center of one structural member to the center of the next member.

**Overhang** — The lower part of a roof that projects over the exterior wall. It is also the horizontal distance which a rafter extends beyond a wall.

**Particle Board** — Wood chips with a resin binder pressed into sheets of various thicknesses.

**Penny** — A term used to indicate the length of nail. The letter "d" is used as an abbreviation for the term penny.

**Perimeter** — The length of the outer boundary of a plane figure.

**Pier** — A piece of solid upright masonry designed to support a structural load.

**Pitch (Roof)** — The incline of a roof expressed as rise divided by the span.

**Plate** — A horizontal structural member to which the top and bottom of the studs are secured.

**Plywood** — Boards made of several layers of wood. The grain of alternate pieces is at right angles.

**Potter's Wheel** — A horizontal wheel activated by the potter's feet, thus rotating his clay.

**Projection** — The line length of rafter which extends beyond the plate.

**Radius** — Distance from the center of a circle to any point on the circumference.

**Rafter** — Main structural roof support.

**Railing (Stair)** — A complete assembly consisting of handrail, balustrade, and newel post.

**Retaining Wall** — Concrete or masonry wall built to support the earth in a sloping area.

**Ribbon** — A narrow board secured to the edge of studs to support the joists.

**Ridge** — The uppermost horizontal portion of a roof.

**Right Triangle** — A triangle with a 90° angle.

**Rise** — The change in elevation of a roof or set of stairs.

**Roof Boards** — Boards which cover the exterior of the roof rafters.

**Run** — The distance from the face of the upper stair riser to the face of the lower stair riser. It is also the horizontal distance covered by a rafter.

**Saw Kerf** — Groove in wood to prevent warping, relieve stress or binding.

**Scaffold** — A temporary platform used to support workers and equipment.

**Semicircular** — One-half of a circle.

**Septic Tank** — An underground metal, masonry, or wooden tank for retaining solid sewage which is then decomposed by bacteria.

**Sheathing** — Lumber or man-made sheets placed on the exterior of the studs or rafters.

**Shim** — A thin often tapered strip of wood or metal used to fill in a space when leveling.

**Shingle** — Wood, asphalt, or asbestos units which are applied to the roof or sidewalls to make the surface weathertight.

**Shoe Mold** — The molding which conceals the joint between the floor and baseboard.

**Siding** — The exterior covering of the sides of a building.

**Sill** — The lowest horizontal member of a frame structure, usually resting upon a foundation. It is also the lowest member of a door or window frame.

**Slope** — The incline of a roof expressed as a ratio of rise to run.

**Soffit** — The underside of the members of a building, such as the undersurface of the cornice.

**Sole Plate** — The horizontal member on the subfloor to which studs are nailed.

**Span** — The distance between opposite sides of a building.

**Specifications** — A written statement or part of the contract listing the kind and quality of materials or workmanship used in the construction.

**Square** — A quantity of material sufficient to cover one hundred square feet of surface.

**Square Foot** — A unit of area equal to 144 square inches.

**Stair Riser** — The vertical boards located between treads of a set of stairs.

**Stair Tread** — The horizontal boards upon which one steps.

**Story Pole** — A stick or rod used to locate elements in construction. Stair risers, courses of siding, height or location of framing members, and cabinet layout.

**Stringer** — The inclined members of a set of stairs to which the treads and risers are secured, usually made of finished wood.

**Stud** — The vertical members of outside walls or partitions.

**Subfloor** — Plywood or matched boards laid directly on the floor joists upon which a finish floor is laid.

**Tail** — The portion of a rafter which extends beyond a wall.

**Tenon** — A projecting member left by cutting away the wood around it for insertion into a mortise to make a joint.

**Tongue and Groove (T and G)** — Boards that are joined by cutting a tongue (rib) on the edge of one board and a groove on the edge of an adjacent board.

**To The Weather** — A term which refers to the exposed portion of siding or shingle.

**Trapezoid** — A quadrilateral with one and only one pair of sides parallel.

**Trim** — The finish materials in a building such as molding, baseboard, cornice, etc.

**Wallboard** — Gypsum, wood pulp or similar materials made into large sheets which are nailed to the interior of a building frame to form interior walls.

**Window Apron** — Member of the trim of a window located on the wall immediately beneath the stool.

**Window Stool** — A horizontal member placed on top of a window sill on the interior of a window.

# APPENDIX

## SECTION I

## DENOMINATE NUMBERS

Denominate numbers are numbers that include units of measurement. The units of measurement are arranged from the largest units at the left to the smallest unit at the right.

For example: 6 yd. 2 ft. 4 in.

All basic operations of arithmetic can be performed on denominate numbers.

### I. EQUIVALENT MEASURES

Measurements that are equal can be expressed in different terms. For example, 12 in. = 1ft. If these equivalents are divided, the answer is 1.

$$\frac{1 \text{ ft.}}{12 \text{ in.}} = 1 \qquad \frac{12 \text{ in.}}{1 \text{ ft.}} = 1$$

To express one measurement as another equal measurement, multiply by the equivalent in the form of 1.

To express 6 inches in equivalent foot measurement, multiply 6 inches by one in the form of

$\frac{1 \text{ ft.}}{12 \text{ in.}}$. In the numerator and denominator, divide by a common factor.

$$6 \text{ in.} = \frac{\overset{1}{\cancel{6 \text{ in.}}}}{1} \times \frac{1 \text{ ft.}}{\underset{2}{\cancel{12 \text{ in.}}}} = \frac{1}{2} \text{ ft. or } 0.5 \text{ ft.}$$

To express 4 feet in equivalent inch measurement, multiply 4 feet by one in the form of

$\frac{12 \text{ in.}}{1 \text{ ft.}}$.

$$4 \text{ ft.} = \underset{1}{\overset{4}{\cancel{4 \text{ ft.}}}} \times \frac{12 \text{ in.}}{\cancel{1 \text{ ft.}}} = \frac{48 \text{ in.}}{1} = 48 \text{ in.}$$

*Per* means division, as with a fraction bar. For example, 50 miles per hour can be written $\frac{50 \text{ miles}}{1 \text{ hour}}$.

## II.   BASIC OPERATIONS

### A.   ADDITION

SAMPLE:  2 yd. 1 ft. 5 in. + 1 ft. 8 in. + 5 yd. 2 ft.

1.  Write the denominate numbers in a column with like units in the same column.

```
    2 yd.   1 ft.   5 in.
            1 ft.   8 in.
 +  5 yd.   2 ft.
 _____
```

2.  Add the denominate numbers in each column.

```
    7 yd.   4 ft. 13 in.
```

3.  Express the answer using the largest possible units.

```
    7 yd.                  =   7 yd.
            4 ft.          =   1 yd.  1 ft.
                  13 in. = +            1 ft.  1 in.
                          _____
    7 yd.   4 ft. 13 in. =   8 yd.  2 ft.  1 in.
```

### B.   SUBTRACTION

SAMPLE:  4 yd. 3 ft. 5 in. – 2 yd. 1 ft. 7 in.

1.  Write the denominate numbers in columns with like units in the same column.

```
    4 yd.   3 ft.   5 in.
 -  2 yd.   1 ft.   7 in.
 _____
```

2.  Starting at the right, examine each column to compare the numbers. If the bottom number is larger, exchange one unit from the column at the left for its equivalent. Combine like units.

7 in. is larger than 5 in.

3 ft.  = 2 ft. 12 in.
12 in. + 5 in.  = 17 in.

3.  Subtract the denominate numbers.

```
    4 yd.   2 ft. 17 in.
 -  2 yd.   1 ft.  7 in.
 _____
    2 yd.   1 ft. 10 in.
```

4.  Express the answer using the largest possible units.

```
    2 yd.   1 ft. 10 in.
```

### C.   MULTIPLICATION

*— By a constant*

SAMPLE:  1 hr. 24 min. x 3

1.  Multiply the denominate number by the constant.

```
    1 hr.  24 min.
           x 3
 _____
    3 hr.  72 min.
```

2.  Express the answer using the largest possible units.

```
    3 hr.                =  3 hr.
           72 min.       =  1 hr.  12 min.
 _____
    3 hr.  72 min.       =  4 hr.  12 min.
```

*— By a denominate number expressing linear measurement*

SAMPLE:  9 ft. 6 in. x 10 ft.

1. Express all denominate numbers in the same unit.

$$9 \text{ ft. } 6 \text{ in. } = 9\frac{1}{2} \text{ ft.}$$

2. Multiply the denominate numbers. (This includes the units of measure, such as ft. x ft. = sq. ft.)

$$9\frac{1}{2} \text{ ft. x } 10 \text{ ft. } =$$

$$\frac{19}{2} \text{ ft. x } 10 \text{ ft. } =$$

95 sq. ft.

*— By a denominate number expressing square measurement*

SAMPLE:  3 ft. x 6 sq. ft.

1. Multiply the denominate numbers. (This includes the units of measure, such as ft. x ft. = sq. ft. and sq. ft. x ft. = cu. ft.)

3 ft. x 6 sq. ft. = 18 cu. ft.

*— By a denominate number expressing rate*

SAMPLE:  50 miles per hour x 3 hours

1. Express the rate as a fraction using the fraction bar for *per*.

$$\frac{50 \text{ miles}}{1 \text{ hour}} \times \frac{3 \text{ hours}}{1} =$$

2. Divide the numerator and denominator by any common factors, including units of measure.

$$\frac{50 \text{ miles}}{\cancel{1 \text{ hour}}} \times \frac{\cancel{3 \text{ hours}}^{\,3}}{1} =$$

3. Multiply numerators. Multiply denominators.

$$\frac{150 \text{ miles}}{1} =$$

4. Express the answer in the remaining unit.

150 miles

## D.  DIVISION

*— By a constant*

SAMPLE:  8 gal. 3 qt. ÷ 5

1. Express all denominate numbers in the same unit.

8 gal.  3 qt. = 35 qt.

2. Divide the denominate number by the constant.

35 qt. ÷ 5 = 7 qt.

3. Express the answer using the largest possible units.

7 qt. = 1 gal.  3 qt.

*— By a denominate number expressing linear measurement*

SAMPLE: 11 ft. 4 in. ÷ 8 in.

1. Express all denominate numbers in the same unit.

    11 ft. 4 in. = 136 in.

2. Divide the denominate numbers by a common factor. (This includes the units of measure, such as inches ÷ inches = 1.)

    136 in. ÷ 8 in. =

    $$\frac{\overset{17}{\cancel{136}\text{ in.}}}{\underset{1}{\cancel{8}\text{ in.}}} = \frac{17}{1} = 17$$

*— By a linear measure with a square measurement as the dividend*

SAMPLE: 20 sq. ft. ÷ 4 ft.

1. Divide the denominate numbers. (This includes the units of measure, such as sq. ft. ÷ ft. = ft.)

    20 sq. ft. ÷ 4 ft.

    $$\frac{\overset{5 \text{ ft.}}{\cancel{20}\text{ sq. ft.}}}{\cancel{4}\text{ ft.}} = \frac{5 \text{ ft.}}{1}$$

2. Express the answer in the remaining unit.

    5 ft.

*— By denominate numbers used to find rate*

SAMPLE: 200 mi. ÷ 10 gal.

1. Divide the denominate numbers

    $$\frac{\overset{20 \text{ mi.}}{\cancel{200}\text{ mi.}}}{\underset{1 \text{ gal.}}{\cancel{10}\text{ gal.}}} = \frac{20 \text{ mi.}}{1 \text{ gal.}}$$

2. Express the units with the fraction bar meaning *per*.

    $$\frac{20 \text{ mi.}}{1 \text{ gal.}} = 20 \text{ miles per gallon}$$

Note:  Alternate methods of performing the basic operations will produce the same result. The choice of method is determined by the individual.

## SECTION II

### TABLE 1
### STANDARD TABLES OF CUSTOMARY UNITS OF MEASURE

**Linear Measure**

| | | |
|---|---|---|
| 12 inches (in.) | = | 1 foot (ft). |
| 3 ft. | = | 1 yard (yd.) |
| 16 1/2 ft. | = | 1 rod (rd.) |
| 5 1/2 yd. | = | 1 rd. |
| 320 rd. | = | 1 mile |
| 1760 yd. | = | 1 mile |
| 5280 ft. | = | 1 mile |

**Surface Measure**

| | | |
|---|---|---|
| 144 sq. in. | = | 1 sq. ft. |
| 9 sq. ft. | = | 1 sq. yd. |
| 30 1/4 sq. yd. | = | 1 sq. rd. |
| 160 sq. rd. | = | 1 acre |
| 640 acres | = | 1 sq. mile |
| 43,560 sq. ft. | = | 1 acre |

**Cubic Measure**

| | | |
|---|---|---|
| 1728 cu. in. | = | 1 cu. ft. |
| 27 cu. ft. | = | 1 cu. yd. |
| 128 cu. ft. | = | 1 cord |

**Angular (Circular) Measure**

| | | |
|---|---|---|
| 60 sec. ('') | = | 1 min. (') |
| 60' | = | 1 degree (°) |
| 90° | = | 1 quadrant |
| 360° | = | 1 circle |

**Time Measure**

| | | |
|---|---|---|
| 60 seconds (sec.) | = | 1 minute (min.) |
| 60 min. | = | 1 hour (hr.) |
| 24 hr. | = | 1 day |
| 7 days | = | 1 week |
| 52 weeks | = | 1 year |
| 365 days | = | 1 year |
| 10 years | = | 1 decade |

**Liquid Measure**

| | | |
|---|---|---|
| 4 gills | = | 1 pint (pt.) |
| 2 pt. | = | 1 quart (qt.) |
| 4 qt. | = | 1 gallon (gal.) |
| 231 cu. in. | = | 1 gal. |
| 31.5 gal. | = | 1 barrel (bbl.) |
| 42 gal. | = | 1 bbl. of oil |
| 8 1/2 lb. | = | 1 gal. water |
| 7 1/2 gal. | = | 1 cu. ft. |

**Weights of Materials**

| | | |
|---|---|---|
| 0.096 lb. | = | 1 cu. in. aluminum |
| 0.260 lb. | = | 1 cu. in. cast iron |
| 0.283 lb. | = | 1 cu. in. mild steel |
| 0.321 lb. | = | 1 cu. in. copper |
| 0.41 lb. | = | 1 cu. in. lead |
| 112 lb. | = | 1 cu. ft. Dowmetal |
| 167 lb. | = | 1 cu. ft. aluminum |
| 464 lb. | = | 1 cu. ft. cast iron |
| 490 lb. | = | 1 cu. ft. mild steel |
| 555.6 lb. | = | 1 cu. ft. copper |
| 710 lb. | = | 1 cu. ft. lead |

**Avoirdupois Weight**

| | | |
|---|---|---|
| 16 ounces (oz.) | = | 1 pound (lb.) |
| 100 lb. | = | 1 hundredweight (cwt.) |
| 20 cwt. | = | 1 ton |
| 2000 lb. | = | 1 ton |
| 8 1/2 lb. | = | 1 gal. of water |
| 62.4 lb. | = | 1 cu. ft. of water |
| 112 lb. | = | 1 long cwt. |
| 2240 lb. | = | 1 long ton |

**Dry Measure**

| | | |
|---|---|---|
| 2 cups | = | 1 pt. |
| 2 pt. | = | 1 qt. |
| 4 qt. | = | 1 gal. |
| 8 qt. | = | 1 peck (pk.) |
| 4 pk. | = | 1 bushel (bu.) |

**Miscellaneous**

| | | |
|---|---|---|
| 12 units | = | 1 dozen (doz.) |
| 12 doz. | = | 1 gross |
| 144 units | = | 1 gross |
| 24 sheets | = | 1 quire |
| 20 quires | = | 1 ream |
| 20 units | = | 1 score |
| 6 ft. | = | 1 fathom |

# TABLE II

## CONVERSION OF ENGLISH AND METRIC UNITS OF MEASURE

| Linear Measure | | | | | | | | |
|---|---|---|---|---|---|---|---|---|
| Unit | Inches to milli-metres | Milli-metres to inches | Feet to metres | Metres to feet | Yards to metres | Metres to yards | Miles to kilo-metres | Kilo-metres to miles |
| 1 | 25.40 | 0.03937 | 0.3048 | 3.281 | 0.9144 | 1.094 | 1.609 | 0.6214 |
| 2 | 50.80 | 0.07874 | 0.6096 | 6.562 | 1.829 | 2.187 | 3.219 | 1.243 |
| 3 | 76.20 | 0.1181 | 0.9144 | 9.842 | 2.743 | 3.281 | 4.828 | 1.864 |
| 4 | 101.60 | 0.1575 | 1.219 | 13.12 | 3.658 | 4.374 | 6.437 | 2.485 |
| 5 | 127.00 | 0.1968 | 1.524 | 16.40 | 4.572 | 5.468 | 8.047 | 3.107 |
| 6 | 152.40 | 0.2362 | 1.829 | 19.68 | 5.486 | 6.562 | 9.656 | 3.728 |
| 7 | 177.80 | 0.2756 | 2.134 | 22.97 | 6.401 | 7.655 | 11.27 | 4.350 |
| 8 | 203.20 | 0.3150 | 2.438 | 26.25 | 7.315 | 8.749 | 12.87 | 4.971 |
| 9 | 228.60 | 0.3543 | 2.743 | 29.53 | 8.230 | 9.842 | 14.48 | 5.592 |

Example   1 in. = 25.40 mm,   1 m = 3.281 ft.,   1 km = 0.6214 mi.

| Surface Measure | | | | | | | | | | |
|---|---|---|---|---|---|---|---|---|---|---|
| Unit | Square inches to square centi-metres | Square centi-metres to square inches | Square feet to square metres | Square metres to square feet | Square yards to square metres | Square metres to square yards | Acres to hec-tares | Hec-tares to acres | Square miles to square kilo-metres | Square kilo-metres to square miles |
| 1 | 6.452 | 0.1550 | 0.0929 | 10.76 | 0.8361 | 1.196 | 0.4047 | 2.471 | 2.59 | 0.3861 |
| 2 | 12.90 | 0.31 | 0.1859 | 21.53 | 1.672 | 2.392 | 0.8094 | 4.942 | 5.18 | 0.7722 |
| 3 | 19.356 | 0.465 | 0.2787 | 32.29 | 2.508 | 3.588 | 1.214 | 7.413 | 7.77 | 1.158 |
| 4 | 25.81 | 0.62 | 0.3716 | 43.06 | 3.345 | 4.784 | 1.619 | 9.884 | 10.36 | 1.544 |
| 5 | 32.26 | 0.775 | 0.4645 | 53.82 | 4.181 | 5.98 | 2.023 | 12.355 | 12.95 | 1.931 |
| 6 | 38.71 | 0.93 | 0.5574 | 64.58 | 5.017 | 7.176 | 2.428 | 14.826 | 15.54 | 2.317 |
| 7 | 45.16 | 1.085 | 0.6503 | 75.35 | 5.853 | 8.372 | 2.833 | 17.297 | 18.13 | 2.703 |
| 8 | 51.61 | 1.24 | 0.7432 | 86.11 | 6.689 | 9.568 | 3.237 | 19.768 | 20.72 | 3.089 |
| 9 | 58.08 | 1.395 | 0.8361 | 96.87 | 7.525 | 10.764 | 3.642 | 22.239 | 23.31 | 3.475 |

Example   1 sq. in. = 6.452 cm$^2$,   1 m$^2$ = 1.196 sq. yd.,   1 sq. mi. = 2.59 km$^2$

| Cubic Measure | | | | | | | | |
|---|---|---|---|---|---|---|---|---|
| Unit | Cubic inches to cubic centi-metres | Cubic centi-metres to cubic inches | Cubic feet to cubic metres | Cubic metres to cubic feet | Cubic yards to cubic metres | Cubic metres to cubic yards | Gallons to cubic feet | Cubic feet to gallons |
| 1 | 16.39 | 0.06102 | 0.02832 | 35.31 | 0.7646 | 1.308 | 0.1337 | 7.481 |
| 2 | 32.77 | 0.1220 | 0.05663 | 70.63 | 1.529 | 2.616 | 0.2674 | 14.96 |
| 3 | 49.16 | 0.1831 | 0.08495 | 105.9 | 2.294 | 3.924 | 0.4010 | 22.44 |
| 4 | 65.55 | 0.2441 | 0.1133 | 141.3 | 3.058 | 5.232 | 0.5347 | 29.92 |
| 5 | 81.94 | 0.3051 | 0.1416 | 176.6 | 3.823 | 6.540 | 0.6684 | 37.40 |
| 6 | 98.32 | 0.3661 | 0.1699 | 211.9 | 4.587 | 7.848 | 0.8021 | 44.88 |
| 7 | 114.7 | 0.4272 | 0.1982 | 247.2 | 5.352 | 9.156 | 0.9358 | 52.36 |
| 8 | 131.1 | 0.4882 | 0.2265 | 282.5 | 6.116 | 10.46 | 1.069 | 59.84 |
| 9 | 147.5 | 0.5492 | 0.2549 | 371.8 | 6.881 | 11.77 | 1.203 | 67.32 |

Example   1 cm$^3$ = 0.06102 cu. in.,   1 gal. = 0.1337 cu. ft.

| Volume or Capacity Measure | | | | | | | | | | |
|---|---|---|---|---|---|---|---|---|---|---|
| Unit | Liquid ounces to cubic centi-metres | Cubic centi-metres to liquid ounces | Pints to litres | Litres to pints | Quarts to litres | Litres to quarts | Gallons to litres | Litres to gallons | Bushels to hecto-litres | Hecto-litres to bushels |
| 1 | 29.57 | 0.03381 | 0.4732 | 2.113 | 0.9463 | 1.057 | 3.785 | 0.2642 | 0.3524 | 2.838 |
| 2 | 59.15 | 0.06763 | 0.9463 | 4.227 | 1.893 | 2.113 | 7.571 | 0.5284 | 0.7048 | 5.676 |
| 3 | 88.72 | 0.1014 | 1.420 | 6.340 | 2.839 | 3.785 | 11.36 | 0.7925 | 1.057 | 8.513 |
| 4 | 118.3 | 0.1353 | 1.893 | 8.454 | 3.170 | 4.227 | 15.14 | 1.057 | 1.410 | 11.35 |
| 5 | 147.9 | 0.1691 | 2.366 | 10.57 | 4.732 | 5.284 | 18.93 | 1.321 | 1.762 | 14.19 |
| 6 | 177.4 | 0.2029 | 2.839 | 12.68 | 5.678 | 6.340 | 22.71 | 1.585 | 2.114 | 17.03 |
| 7 | 207.0 | 0.2367 | 3.312 | 14.79 | 6.624 | 7.397 | 26.50 | 1.849 | 2.467 | 19.86 |
| 8 | 236.6 | 0.2705 | 3.785 | 16.91 | 7.571 | 8.454 | 30.28 | 2.113 | 2.819 | 22.70 |
| 9 | 266.2 | 0.3043 | 4.259 | 19.02 | 8.517 | 9.510 | 34.07 | 2.378 | 3.171 | 25.54 |

Example   1 L = 2.113 pt.,   1 gal. = 3.785 L

## METRIC AND CUSTOMARY DECIMAL EQUIVALENTS
## FOR FRACTIONAL PARTS OF AN INCH

| Fraction | Decimal Equivalent | | Fraction | Decimal Equivalent | |
|---|---|---|---|---|---|
| | Customary (in.) | Metric (mm) | | Customary (in.) | Metric (mm) |
| 1/64— | 0.015625 | 0.3969 | 33/64— | 0.515625 | 13.0969 |
| 1/32— | 0.03125 | 0.7938 | 17/32— | 0.53125 | 13.4938 |
| 3/64— | 0.046875 | 1.1906 | 35/64— | 0.546875 | 13.8906 |
| 1/16— | 0.0625 | 1.5875 | 9/16— | 0.5625 | 14.2875 |
| 5/64— | 0.078125 | 1.9844 | 37/64— | 0.578125 | 14.6844 |
| 3/32— | 0.09375 | 2.3813 | 19/32— | 0.59375 | 15.0813 |
| 7/64— | 0.109375 | 2.7781 | 39/64— | 0.609375 | 15.4781 |
| 1/8— | 0.1250 | 3.1750 | 5/8— | 0.6250 | 15.8750 |
| 9/64— | 0.140625 | 3.5719 | 41/64— | 0.640625 | 16.2719 |
| 5/32— | 0.15625 | 3.9688 | 21/32— | 0.65625 | 16.6688 |
| 11/64— | 0.171875 | 4.3656 | 43/64— | 0.671875 | 17.0656 |
| 3/16— | 0.1875 | 4.7625 | 11/16— | 0.6875 | 17.4625 |
| 13/64— | 0.203125 | 5.1594 | 45/64— | 0.703125 | 17.8594 |
| 7/32— | 0.21875 | 5.5563 | 23/32— | 0.71875 | 18.2563 |
| 15/64— | 0.234375 | 5.9531 | 47/64— | 0.734375 | 18.6531 |
| 1/4— | 0.250 | 6.3500 | 3/4— | 0.750 | 19.0500 |
| 17/64— | 0.265625 | 6.7469 | 49/64— | 0.765625 | 19.4469 |
| 9/32— | 0.28125 | 7.1438 | 25/32— | 0.78125 | 19.8438 |
| 19/64— | 0.296875 | 7.5406 | 51/64— | 0.796875 | 20.2406 |
| 5/16— | 0.3125 | 7.9375 | 13/16— | 0.8125 | 20.6375 |
| 21/64— | 0.328125 | 8.3384 | 53/64— | 0.828125 | 21.0344 |
| 11/32— | 0.34375 | 8.7313 | 27/32— | 0.84375 | 21.4313 |
| 23/64— | 0.359375 | 9.1281 | 55/64— | 0.859375 | 21.8281 |
| 3/8— | 0.3750 | 9.5250 | 7/8— | 0.8750 | 22.2250 |
| 25/64— | 0.390625 | 9.9219 | 57/64— | 0.890625 | 22.6219 |
| 13/32— | 0.40625 | 10.3188 | 29/32— | 0.90625 | 23.0188 |
| 27/64— | 0.421875 | 10.7156 | 59/64— | 0.921875 | 23.4156 |
| 7/16— | 0.4375 | 11.1125 | 15/16— | 0.9375 | 23.8125 |
| 29/64— | 0.453125 | 11.5094 | 61/64— | 0.953125 | 24.2094 |
| 15/32— | 0.46875 | 11.9063 | 31/32— | 0.96875 | 24.6063 |
| 31/64— | 0.484375 | 12.3031 | 63/64— | 0.984375 | 25.0031 |
| 1/2— | 0.500 | 12.7000 | 1— | 1.000 | 25.4000 |

## POWERS AND ROOTS OF NUMBERS (1 through 100)

| Num-ber | Powers | | Roots | | Num-ber | Powers | | Roots | |
|---|---|---|---|---|---|---|---|---|---|
| | Square | Cube | Square | Cube | | Square | Cube | Square | Cube |
| 1 | 1 | 1 | 1.000 | 1.000 | 51 | 2,601 | 132,651 | 7.141 | 3.708 |
| 2 | 4 | 8 | 1.414 | 1.260 | 52 | 2,704 | 140,608 | 7.211 | 3.733 |
| 3 | 9 | 27 | 1.732 | 1.442 | 53 | 2,809 | 148,877 | 7.280 | 3.756 |
| 4 | 16 | 64 | 2.000 | 1.587 | 54 | 2,916 | 157,464 | 7.348 | 3.780 |
| 5 | 25 | 125 | 2.236 | 1.710 | 55 | 3,025 | 166,375 | 7.416 | 3.803 |
| 6 | 36 | 216 | 2.449 | 1.817 | 56 | 3,136 | 175,616 | 7.483 | 3.826 |
| 7 | 49 | 343 | 2.646 | 1.913 | 57 | 3,249 | 185,193 | 7.550 | 3.849 |
| 8 | 64 | 512 | 2.828 | 2.000 | 58 | 3,364 | 195,112 | 7.616 | 3.871 |
| 9 | 81 | 729 | 3.000 | 2.080 | 59 | 3,481 | 205,379 | 7.681 | 3.893 |
| 10 | 100 | 1,000 | 3.162 | 2.154 | 60 | 3,600 | 216,000 | 7.746 | 3.915 |
| 11 | 121 | 1,331 | 3.317 | 2.224 | 61 | 3,721 | 226,981 | 7.810 | 3.936 |
| 12 | 144 | 1,728 | 3.464 | 2.289 | 62 | 3,844 | 238,328 | 7.874 | 3.958 |
| 13 | 169 | 2,197 | 3.606 | 2.351 | 63 | 3,969 | 250,047 | 7.937 | 3.979 |
| 14 | 196 | 2,744 | 3.742 | 2.410 | 64 | 4,096 | 262,144 | 8.000 | 4.000 |
| 15 | 225 | 3,375 | 3.873 | 2.466 | 65 | 4,225 | 274,625 | 8.062 | 4.021 |
| 16 | 256 | 4,096 | 4.000 | 2.520 | 66 | 4,356 | 287,496 | 8.124 | 4.041 |
| 17 | 289 | 4,913 | 4.123 | 2.571 | 67 | 4,489 | 300,763 | 8.185 | 4.062 |
| 18 | 324 | 5,832 | 4.243 | 2.621 | 68 | 4,624 | 314,432 | 8.246 | 4.082 |
| 19 | 361 | 6,859 | 4.359 | 2.668 | 69 | 4,761 | 328,509 | 8.307 | 4.102 |
| 20 | 400 | 8,000 | 4.472 | 2.714 | 70 | 4,900 | 343,000 | 8.367 | 4.121 |
| 21 | 441 | 9,261 | 4.583 | 2.759 | 71 | 5,041 | 357,911 | 8.426 | 4.141 |
| 22 | 484 | 10,648 | 4.690 | 2.802 | 72 | 5,184 | 373,248 | 8.485 | 4.160 |
| 23 | 529 | 12,167 | 4.796 | 2.844 | 73 | 5,329 | 389,017 | 8.544 | 4.179 |
| 24 | 576 | 13,824 | 4.899 | 2.884 | 74 | 5,476 | 405,224 | 8.602 | 4.198 |
| 25 | 625 | 15,625 | 5.000 | 2.924 | 75 | 5,625 | 421,875 | 8.660 | 4.217 |
| 26 | 676 | 17,576 | 5.099 | 2.962 | 76 | 5,776 | 438,976 | 8.718 | 4.236 |
| 27 | 729 | 19,683 | 5.196 | 3.000 | 77 | 5,929 | 456,533 | 8.775 | 4.254 |
| 28 | 784 | 21,952 | 5.292 | 3.037 | 78 | 6,084 | 474,552 | 8.832 | 4.273 |
| 29 | 841 | 24,389 | 5.385 | 3.072 | 79 | 6,241 | 493,039 | 8.888 | 4.291 |
| 30 | 900 | 27,000 | 5.477 | 3.107 | 80 | 6,400 | 512,000 | 8.944 | 4.309 |
| 31 | 961 | 29,791 | 5.568 | 3.141 | 81 | 6,561 | 531,441 | 9.000 | 4.327 |
| 32 | 1,024 | 32,798 | 5.657 | 3.175 | 82 | 6,724 | 551,368 | 9.055 | 4.344 |
| 33 | 1,089 | 35,937 | 5.745 | 3.208 | 83 | 6,889 | 571,787 | 9.110 | 4.362 |
| 34 | 1,156 | 39,304 | 5.831 | 3.240 | 84 | 7,056 | 592,704 | 9.165 | 4.380 |
| 35 | 1,225 | 42,875 | 5.916 | 3.271 | 85 | 7,225 | 614,125 | 9.220 | 4.397 |
| 36 | 1,296 | 46,656 | 6.000 | 3.302 | 86 | 7,396 | 636,056 | 9.274 | 4.414 |
| 37 | 1,369 | 50,653 | 6.083 | 3.332 | 87 | 7,569 | 658,503 | 9.327 | 4.481 |
| 38 | 1,444 | 54,872 | 6.164 | 3.362 | 88 | 7,744 | 681,472 | 9.381 | 4.448 |
| 39 | 1,521 | 59,319 | 6.245 | 3.391 | 89 | 7,921 | 704,969 | 9.434 | 4.465 |
| 40 | 1,600 | 64,000 | 6.325 | 3.420 | 90 | 8,100 | 729,000 | 9.487 | 4.481 |
| 41 | 1,681 | 68,921 | 6.403 | 3.448 | 91 | 8,281 | 753,571 | 9.539 | 4.498 |
| 42 | 1,764 | 74,088 | 6.481 | 3.476 | 92 | 8,464 | 778,688 | 9.592 | 4.514 |
| 43 | 1,849 | 79,507 | 6.557 | 3.503 | 93 | 8,649 | 804,357 | 9.644 | 4.531 |
| 44 | 1,936 | 85,184 | 6.633 | 3.530 | 94 | 8,836 | 830,584 | 9.695 | 4.547 |
| 45 | 2,025 | 91,125 | 6.708 | 3.557 | 95 | 9,025 | 857,375 | 9.747 | 4.563 |
| 46 | 2,116 | 97,336 | 6.782 | 3.583 | 96 | 9,216 | 884,736 | 9.798 | 4.579 |
| 47 | 2,209 | 103,823 | 6.856 | 3.609 | 97 | 9,409 | 912,673 | 9.849 | 4.595 |
| 48 | 2,304 | 110,592 | 6.928 | 3.634 | 98 | 9,604 | 941,192 | 9.900 | 4.610 |
| 49 | 2,401 | 117,649 | 7.000 | 3.659 | 99 | 9,801 | 970,299 | 9.950 | 4.626 |
| 50 | 2,500 | 125,000 | 7.071 | 3.684 | 100 | 10,000 | 1,000,000 | 10.000 | 4.642 |

# ANSWERS TO ODD-NUMBERED PROBLEMS

## SECTION 1

### UNIT 1 ADDITION OF WHOLE NUMBERS

1. 112 feet
3. 7,088
5. $3,749
7. 4,186 square inches
9. 57,474 board feet
11. 33 feet

13. 4,510 square feet
15. $7,312
17. 8,085 board feet
19. 510 square feet
21. $771
23. 3,498 board feet

25. 1,908 board feet
27. a. 150 feet
    b. 448 feet
    c. 276 feet
    d. 330 feet
    e. 552 feet
    f. 264 feet

27. g. 226 feet
    h. 190 feet
    i. 532 feet
    j. 460 feet
    k. 300 feet
    l. 292 feet
29. 174 feet

### UNIT 2 SUBTRACTION OF WHOLE NUMBERS

1. 77 centimetres
3. 269 ounces
5. 2,486 square feet
7. 5,520 gallons

9. 13,053
11. 20,900 board feet
13. 340 millimetres
15. 2,766 feet

17. 475 cubic metres
19. $715
21. $855.00
23. 5'-0"
25. 15'-1"

27. 9'-6"
29. 14'-0"
31. 871 sheets
33. 7'-11'
35. 15'-7"

### UNIT 3 MULTIPLICATION OF WHOLE NUMBERS

1. 704 centimetres
3. 4,064
5. 112,158 inches
7. 139,708 yards
9. $68,685
11. 1,077,249 gallons

13. 175 feet
15. 1,600 feet
17. $784.00
19. 2,340 square feet
21. 64,500 shingles
23. 2,279 board feet

25. 1,080 feet
27. a. A = 22 feet
    b. Section I = 528 feet
       Section II = 1,725 feet
    c. Total = 2,253 square feet
29. 2,380 square feet

### UNIT 4 DIVISION OF WHOLE NUMBERS

1. 141 inches
3. 7,776, R1
5. 101 centimetres

7. 28, R 92
9. 1 054 millimetres
11. 11 hours

13. 21 hours
15. 49 studs
17. 7 inches
19. 74 rafters

21. 37 floor joists
23. 46 joists
25. 117 joists
27. 90 piers

## SECTION 2 COMMON FRACTIONS

### UNIT 5 ADDITION OF COMMON FRACTIONS

1. 1 3/8 inches
3. 1 1/4 gallons
5. 2 7/24 square feet
7. 7 1/4 hours
9. 13/16 inch

11. 1 3/4 inches
13. 28 inches
15. 12 1/4 inches
17. 4 inches
19. 5 1/8 inches

21. 2 inches
23. 4 17/32 inches
25. 1 ft. 1/4 in.
27. 2 ft. 6 1/2 in.
29. 2 ft. 7 1/2 in.

31. 10 1/16 inches
33. 10 7/8 inches
35. 3 ft. 4 in.
37. 41 ft. 3 3/4 in.
39. 54 ft. 1 9/16 in.

### UNIT 6 SUBTRACTION OF COMMON FRACTIONS

1. 3/8 gallon
3. 3/8 board foot
5. 4/9
7. 7/8 metre
9. 9/32 inch
11. 13/16 inch

13. 3/4 inch
15. 1/4 inch
17. 1 inch
19. 2'-4 3/4"
21. 4 1/2 in.
23. 3'-0 1/2"

25. 7/8 inch
27. 3'-5 1/4"
29. 10 1/4"
31. 1/8 inch
33. 3/4 inch
35. 4'-10 3/8"

## UNIT 7 MULTIPLICATION OF COMMON FRACTIONS

| | | | | | | | |
|---|---|---|---|---|---|---|---|
| 1. | 15/64 inch | 13. | 11'-1'' | 27. | 10'-0 1/4'' | 33. | a. 7'-1 9/16'' |
| 3. | 10 3/32 gallons | 15. | 19'-1 1/2'' | 29. | A = 38'' | | b. 10'-1 7/8'' |
| 5. | 37 8/9 | 17. | 6'-3'' | | B = 42 3/4'' | | c. 3'-1 15/16'' |
| 7. | 3/32 board foot | 19. | 1 1/4 inches | | C = 80 3/4'' | | d. 7'-8 1/8'' |
| 9. | 7/32 | 21. | 4 1/2 inches | | D = 23 3/4'' | | e. 8'-2 7/8'' |
| 11. | 11'-5 3/4'' | 23. | 11 7/8 inches | | E = 57'' | | f. 1'-3 11/16'' |
| | | 25. | 21 7/8 inches | 31. | 709 square feet | 35. | 8'-9 3/4'' |

## UNIT 8 DIVISION OF COMMON FRACTIONS

| | | | | | | | |
|---|---|---|---|---|---|---|---|
| 1. | 1/4 | 11. | 40 pieces | 21. | 4 strips | 31. | 52 feet |
| 3. | 4 yards | 13. | 8 metres | 23. | 18 columns | 33. | 13 risers |
| 5. | 5/6 gallon | 15. | 80 table tops | 25. | 40 1/2' | 35. | 1 1/4 inches |
| 7. | 2 1/3 | 17. | 10 pieces | 27. | 5 1/4 inches | 37. | 22 boards |
| 9. | 1 10/33 inches | 19. | 4'-2'' | 29. | 16 spaces | 39. | 16 risers |
| | | | | | | 41. | 16 risers |

## UNIT 9 PRACTICE WITH COMMON FRACTIONS

| | | | | | | | |
|---|---|---|---|---|---|---|---|
| 1. | 5/16 inch | 21. L. | 1 1/16 inches | 29. | 3/4 inch | 53. | 6 13/23 inches |
| 3. | 7/8 inch | M. | 1 5/8 inches | 31. | 15/16 pound | 55. | 17 1/2 |
| 5. | 300 | N. | 2 3/16 inches | 33. | 2 7/16 | 57. | 9 1/3 |
| 7. | 6 | O. | 2 13/16 inches | 35. | 4 1/16 inches | 59. | 3/4 square inch |
| 9. | 40 | P. | 3 5/16 inches | 37. | 5/8 inch | 61. | 2'-8'' |
| 11. | 10 | Q. | 3 15/16 inches | 39. | 5/16 board foot | 63. | 3 3/4 in. |
| 13. | 4 | R. | 4 13/16 inches | 41. | 1 3/16 inches | 65. | 2'-7 3/4'' |
| 15. | 6 | S. | 5 3/8 inches | 43. | 3 3/16 inches | 67. | 18 1/3 |
| 17. | 9/16 inch | T. | 5 11/16 inches | 45. | 1 | 69. | 92 4/7 |
| 19. | 2 7/16 inches | 23. | 1 1/4 inches | 47. | 4 3/8 inches | 71. | 1/3 foot |
| 21. J. | 5/16 inch | 25. | 4 3/10 cm | 49. | 84 cm | 73. | 5 3/8 inches |
| K. | 3/4 inch | 27. | 3/8 | 51. | 29/32 inch | 75. | 2'-10 1/4'' |

## SECTION 3 DECIMAL FRACTIONS

## UNIT 10 ADDITION OF DECIMAL FRACTIONS

| | | | | | | | |
|---|---|---|---|---|---|---|---|
| 1. | 2.3 | 9. | 1.63 inches | 19. | $240.75 | 29. | 694.06 board feet |
| 3. | 184.074 | 11. | 1.9195 inches | 21. | $3,454.94 | 31. | $17.35 |
| 5. | 29.568 | 13. | 0.625 inch | 23. | $127.70 | 33. | $83.10 |
| 7. | 1.5 inches | 15. | $2,056.99 | 25. | $2,477.65 | 35. | $3,000.77 |
| | | 17. | $151.20 | 27. | $120.70 | 37. | 54.6875 inches |

## UNIT 11 SUBTRACTION OF DECIMAL FRACTIONS

| | | | | | | | |
|---|---|---|---|---|---|---|---|
| 1. | $2,188.28 | 9. | 2.13 km | 17. | $728.46 | 27. | $16.20 |
| 3. | $0.55 | 11. | 0.035 inch | 19. | $64.80 | 29. | $372.30 |
| 5. | 0.139 | 13. | 0.24 pound | 21. | $19.50 | 31. | $137.50 |
| 7. | 1.30 litres | 15. | 2.125 inches | 23. | 7 hours | 33. | 88.50 inches |
| | | | | 25. | 3.1875 inches | 35. | $2,700.52 |

## UNIT 12 MULTIPLICATION OF DECIMAL FRACTIONS

| | | | | | | | |
|---|---|---|---|---|---|---|---|
| 1. | 5.162 5 m | 7. | 0.003788 foot | 13. | 640.8 pounds | 19. | 37.44 hours |
| 3. | 6.057 49 g | 9. | 46.7908 | 15. | 888.75 pounds | 21. | 9.75 hours |
| 5. | 2,218.557 | 11. | 2152.8 miles | 17. | 206.25 firebricks | 23. | 1,987.5 pounds |
| | | | | | | 25. | 48.96 hours |

## UNIT 13 DIVISION OF DECIMAL FRACTIONS

| | | | | | | | |
|---|---|---|---|---|---|---|---|
| 1. | 1.233 inches | 7. | 40 | 13. | 64,000 | 19. | 68 shims |
| 3. | 5.638 | 9. | 50 | 15. | 4.5 inches | 21. | $3.69 |
| 5. | $27.32 | 11. | 6.4 | 17. | 24 table tops | 23. | 12 treads |
| | | | | | | 25. | 21 courses |

## UNIT 14 EXPRESSING COMMON FRACTIONS AND MIXED NUMBERS AS DECIMALS

| | | | | | | | |
|---|---|---|---|---|---|---|---|
| 1. | 0.375 | 17. | 127/1000 inch | 25. | a. 0.78125 in. | 37. | 21'-3 3/8'' |
| 3. | 0.25 | 19. | 631/1000 inch | | b. 0.065 ft. | 39. | 13'-4 15/16'' |
| 5. | 0.875 | 21. | a. 0.042 ft. | | c. 157.058 ft. | 41. | 76'-4 1/2'' |
| 7. | 25/32 | | b. 0.104 ft. | 27. | 141.413 ft. | 43. | 7'-0 27/32'' |
| 9. | 17/32 | | c. 127.846 ft. | 29. | 146.003 ft. | | |
| 11. | 1.25 pounds | 23. | a. 0.75 in. | 31. | 5/8 | | |
| 13. | 0.125 inch | | b. 0.063 ft. | 33. | 2 3/4 | | |
| 15. | 0.4375 inch | | c. 113.513 ft. | 35. | 2 | | |

## SECTION 4 PERCENT AND PERCENTAGE

## UNIT 15 SIMPLE PERCENT AND PERCENTAGE

| | | | | | | | |
|---|---|---|---|---|---|---|---|
| 1. | a. 580 | 7. | 119.4 pounds | 19. | a. $22,795 labor | 27. | 15% |
| | b. 68.75 m | 9. | 1,407 bricks | | b. $25,705 material | 29. | $55,485 |
| | c. 70% | 11. | 145.6 board feet | 21. | 1.73% | 31. | $47,995.44 |
| | d. 91.696 | 13. | $33,700 | 23. | 1,650 board feet | 33. | 22% |
| | e. 125% | 15. | $15,000 | 25. | a. $6.88 profit | | |
| 3. | 3,060 board feet | 17. | 49.2% | | b. $75.68 charge | | |
| 5. | 1,350 square feet | | | | | | |

## UNIT 16 INTEREST

| | | | | | |
|---|---|---|---|---|---|
| 1. | $108 interest; $408 amount | 9. | $5,850 | 17. | $526.06 |
| 3. | $35.20 interest; $675.20 amount | 11. | $552.50 | 19. | $679.50 |
| 5. | $128.40 interest; $4,408.40 amount | 13. | $3,750 | 21. | $225.12 |
| 7. | $3,100 interest; $23,100 amount | 15. | 14% | 23. | $4,483.67 |

## UNIT 17 DISCOUNT

| | | | | | | | |
|---|---|---|---|---|---|---|---|
| 1. | $4,989.60 | 15. | $33.88 | 25. | $5.78 | 37. | a. Ready-made is |
| 3. | $4,984.67 | 17. | $7,000 | 27. | $14.25 | | cheaper |
| 5. | $26.46 | 19. | $715 | 29. | $631.14 | | b. By $0.28 |
| 7. | $790.11 | 21. | $181.18 | 31. | $324 | 39. | $508.96 |
| 9. | 20% | 23. | a. $15.85 amount | 33. | $146.68 | 41. | $639.94 |
| 11. | $1,250 | | b. 2% rate of | 35. | $69.04 | 43. | $30.54 |
| 13. | $175 | | discount | | | | |

## SECTION 5 MEASUREMENT: DIRECT AND COMPUTED

## UNIT 18 LINEAR MEASURE

| | | | | | | | | | |
|---|---|---|---|---|---|---|---|---|---|
| 1. | a. 3 1/8'' | 3. | a. 84 mm | 11. | 9 ft. 8 in. | 27. | 40 ft. 10 3/32 in. | 41. | 41 ft. |
| | b. 2 3/8'' | | b. 12 cm | 13. | 58 ft. | 29. | 70 ft. 1 15/16 in. | 43. | 43 ft. |
| | c. 4 5/8'' | | c. 2.5 cm | 15. | 50 ft. | 31. | 56.549 m | 45. | 82 ft. |
| | d. 4 1/8'' | | d. 73 mm | 17. | 740 mm | 33. | 91.106 ft. | 47. | 76 ft. |
| | e. 3 1/4'' | | e. 104 mm | 19. | 86 mm | 35. | 70 ft. | 49. | 10.283 m |
| | f. 1 3/8'' | 5. | 72 in. | 21. | 41 ft. 10 in. | 37. | 62 ft. | 51. | 58 ft. |
| | g. 4 1/2'' | 7. | 100 m | 23. | 45.04 m | 39. | 25 ft. | 53. | 3,912 ft. |
| | h. 3 9/16'' | 9. | 168 km | 25. | 8 ft. 4 17/32 in. | | | | |
| | i. 4 1/8'' | | | | | | | | |

## UNIT 19  SQUARE MEASURE

| | | |
|---|---|---|
| 1. 121 square feet | 17. 904 square feet | 31. 6,863.5 square feet |
| 3. 784 square metres | 108 square inches | 33. $9.00 |
| 5. 841 square centimetres | 19. 280 square feet | 35. 1,338.75 square feet |
| 7. 650 square feet | 72 square inches | 37. 113 1/4 square feet |
| 36 square inches | 21. 238 square feet | 39. 20 square metres |
| 9. 12 square feet | 23. 56 square metres | 41. 28.463 square yards |
| 36 square inches | 25. 232.5 square feet | 43. 30.458 square feet |
| 11. 35 square feet | 27. 420 square feet | 45. 175 square feet |
| 13. 153 square metres | 36 square inches | 47. 2,514 5/9 square yards |
| 15. 1 760 square centimetres | 29. 95.7 square metres | |

## UNIT 20  SURFACE MEASUREMENT — TRIANGLES

| | | |
|---|---|---|
| 1. 72 square feet | 11. 4 square feet | 17. 360 square feet |
| 3. 10 square feet | 116.8 square inches | 19. 181.5 square feet |
| 55 square inches | 13. a. 12 square feet | 21. a. 84.968 square feet |
| 5. 250 square feet | b. 60.5 square feet | b. 254.904 square feet |
| 7. 27 1/8 square feet | c. 260.5 square feet | 23. 2,000 square feet |
| 9. 1,224 square feet | 15. 150 square feet | |

## UNIT 21  THE FRAMING SQUARE

| | | |
|---|---|---|
| 1. 9 feet 1 3/8 inches | 3. 15 feet 7 7/8 inches | 5. 12 inches rise per foot of run |

## UNIT 22  SURFACE MEASUREMENT — IRREGULAR FIGURES

| | | |
|---|---|---|
| 1. 59.25 square feet | 5. 385.5 square feet | 11. 18 2/3 square feet |
| 3. a. 363 61/72 square feet | 7. 4,137.5 square inches | 13. 38 1/3 square yards |
| b. 148 square feet | 9. 23 2/3 square feet | |

## UNIT 23  SURFACE MEASUREMENT — CIRCLES

| | | |
|---|---|---|
| 1. 7 square feet 10 square inches | 7. 198 square inches | 13. 2 146 square centimetres |
| 3. 113 square feet 14 square inches | 9. 27.271 square yards | 15. 899.78 square feet |
| 5. 153.938 square centimetres | 11. 4.909 square yards | |

## UNIT 24  VOLUME MEASUREMENT — CUBES AND RECTANGULAR SOLIDS

| | | |
|---|---|---|
| 1. 729 cubic feet | 11. 56 cubic feet | 23. 360 cubic feet |
| 3. 125 cubic centimetres | 13. .55 cubic yards | 25. 112.5 cubic feet |
| 5. 708 cubic feet | 15. 1 13/27 cubic yards | 27. 6.9444 cubic yards |
| 1,619 cubic inches | 17. 3,792 16/27 cubic yards | 29. 103.5 cubic feet |
| 7. 120 cubic metres | 19. 280 cubic yards | 31. 30.2 cubic yards |
| 9. 200 cubic feet | 21. 354 2/3 cubic yards | |

## UNIT 25  BOARD MEASURE

| | | | |
|---|---|---|---|
| 1. 45 board feet | 11. 96 board feet | 21. 10 board feet | 29. 18 2/3 board feet |
| 3. 36 board feet | 13. 1,120 board feet | 23. 125 board feet | 31. 112 board feet |
| 5. 362 2/3 board feet | 15. 140 board feet | 25. 18 board feet | 33. 640 board feet |
| 7. 280 board feet | 17. 140 board feet | 27. 160 board feet | 35. 1,468.32 board feet |
| 9. 968 board feet | 19. 140 board feet | | |

## UNIT 26 VOLUME MEASUREMENT — CYLINDERS

| | | |
|---|---|---|
| 1. 12.57 cubic feet | 7. 5.94 cubic yards | 13. 60.13 cubic feet |
| 3. 28,202.77 gallons | 9. 4.19 cubic feet | 15. 62.48 cubic feet |
| 5. 45.74 cubic yards | 11. 23.97 cubic feet | 17. 1,130.98 square feet |

## UNIT 27 WEIGHT MEASURE

| | | | |
|---|---|---|---|
| 1. 1,293 ounces | 9. 56,000 pounds | 17. $15.50 per box | 23. 10.10625 tons |
| 3. 310,300 pounds | 11. 960 pounds | 19. 300 pounds | 25. 1,209.6 pounds |
| 5. 0.3 kilogram | 13. 121.925 pounds | 21. 129 3/8 pounds | 27. 21.66 pounds |
| 7. 24 square feet | 15. 21 1/3 pounds | | |

## SECTION 6 RATIO AND PROPORTION
### UNIT 28 RATIO

| | | | | |
|---|---|---|---|---|
| 1. 3:5 | 9. 1:40 | 17. 3/4 | 25. 1/8 | 31. 5/12 |
| 3. 1:6 | 11. 1:10 | 19. 7/50 | 27. 1/8 | 33. 1/6 |
| 5. 12:5 | 13. 3/2 | 21. 1/3 | 29. 1/3 | 35. 1/4 |
| 7. 8:1 | 15. 3/4 | 23. 5/8 | | |

## UNIT 29 PROPORTION

| | | | |
|---|---|---|---|
| 1. 16 | 9. 60 | 17. 27 feet | 25. 5/98 |
| 3. 7 | 11. $1 | 19. 36 days | 27. 270 revolutions |
| 5. 440 cubic yards | 13. 64 | 21. 14.816 pounds | per minute |
| 7. 16 2/3 | 15. 6 centimetres | 23. 4 2/3 gallons | 29. 2 inches |

## SECTION 7 POWERS AND ROOTS
### UNIT 30 APPLICATION OF EXPONENTS IN FORMULAS

| | | |
|---|---|---|
| 1. 64 square inches | 17. 3.1416 square metres | 29. 256 square feet |
| 3. 4 square inches | 19. 72.3825 square inches | 31. 2 square feet |
| 5. 36 square inches | 21. 2 cubic feet | 36 square inches |
| 7. 1 square foot | 1,369.5 cubic inches | 33. 576 square feet |
| 9. 9 square feet | 23. 359 cubic feet | 35. 804 square feet |
| 11. 6.25 square inches | 327.63 cubic inches | 35.94 square inches |
| 13. 52 cubic feet | 25. 3 cubic feet | 37. 1,520 square feet |
| 1,269 cubic inches | 357.78 cubic inches | 76.95 square inches |
| 15. 20 square feet | 27. 13 cubic feet | 39. 8 cubic feet |
| 139.08 square inches | 1,663.48 cubic inches | |

## UNIT 31 USING SQUARE ROOT TO FIND SIDES OF RIGHT TRIANGLES

| | | |
|---|---|---|
| 1. 9 | 15. 27.66 centimetres | 29. 54 feet 3 19/32 inches |
| 3. 13 | 17. 13 feet 0 15/32 inches | 31. 58 feet 8 9/32 inches |
| 5. 23 | 19. 7 feet 8 1/32 inches | 33. 17 feet 1 27/32 inches |
| 7. 29.87 | 21. 15 feet 3 3/16 inches | 35. 9 feet 7 11/32 inches |
| 9. 41.13 | 23. 23 feet 0 3/8 inches | 37. 18 feet 10 11/16 inches |
| 11. 17 1/32 inches | 25. 20 feet | 39. 20 feet 1 1/2 inches |
| 13. 8 15/16 inches | 27. 4 feet 6 21/32 inches | |

## SECTION 8  ESTIMATING
### UNIT 32  GIRDERS

1.   120 board feet
3.   120 board feet

5.   230 board feet
7.   176 board feet

9.   a. 4 Lally columns
     b. $21

### UNIT 33  SILLS

1.   24 board feet
3.   a. 224 board feet
     b. 14 pieces

5.   560 board feet
7.   a. 288 feet
     b. 288 board feet

### UNIT 34  FLOOR JOISTS

1.   1,269 1/3 board feet   3.   917 1/3 board feet   5.   a. 938 2/3 board feet   b. 28 feet

### UNIT 35  BRIDGING

1.   72 linear feet   3.   108 linear feet   5.   64 board feet   7.   360 linear feet

### UNIT 36  ROUGH FLOORING

1.   5 board feet   3.   1,664 board feet   5.   5,760 board feet   7.   1,344 square feet

### UNIT 37  WALL PLATES

1.   16 board feet   3.   320 board feet   5.   25 pieces

### UNIT 38  STUDDING AND FIRE OR DRAFT STOPS

1.   8 linear feet   3.   128 studs   5.   1,056 linear feet   7.   157 pieces

### UNIT 39  RAFTERS

1.   a. 7 ft. 9 15/16 in.
     b. 352 bd. ft.

3.   1,360 bd. ft.   5.   48 ft.   7.   48 feet

### UNIT 40  SHEATHING

1.   270.4 square feet   3.   38 sheets   5.   2,880 square feet

### UNIT 41  TRIM

1.   27 feet   3.   26 feet

### UNIT 42  ROOFING

1.   35 bundles   3.   352.5 pounds   5.   8.316 squares   7.   2 rolls

### UNIT 43  DOORS AND WINDOWS

1.   $269.76   3.   $8.11   5.   $35.02

### UNIT 44  SIDING

1.   2,655 square feet   3.   2,726 square feet   5.   720 square feet   7.   135 square feet

## UNIT 45  STAIRS AND INTERIOR DOOR JAMBS

1.  11 risers
3.  7 inch rise, 10 1/2 inch tread
5.  16 risers
7.  4 1/2 inches

## UNIT 46  FINISH FLOORING AND PAPER

1.  1,793.8 board feet
3.  682.325 board feet
5.  1,080.8 square feet

## UNIT 47  HARDWARE AND SUPPLIES

1.  75 pounds
3.  20 bolts
5.  5.52 or 6 gallons

## ACHIEVEMENT REVIEW A

1.   621 square feet
3.   81 feet 3 inches
5.   52.6875 millimetres
7.   3.5 hours
9.   0.8125 inch
11.  $41,808
13.  0.875
15.  $31,050
17.  $8.99
19.  17 square feet
21.  255 cubic feet
23.  2 inches
25.  4 cubic feet
27.  A: 104 square metres
     B: 248 square inches
29.  21.65 square centimetres
31.  398.67 cubic metres
33.  $18.40

## ACHIEVEMENT REVIEW B

1.   $9,038
3.   504 feet
5.   37 feet 8 1/4 inches
7.   8 feet 3 3/4 inches
9.   $5,415.78
11.  $80,640
13.  0.375
15.  $7,392
17.  $1,888.36
19.  9.75 square feet
21.  896 cubic metres
23.  a. 3:5
     b. 1:6
     c. 1:2
25.  64 cubic feet
27.  A: 92 square inches
     B: 170 square centimetres
29.  28.274 square inches
31.  0.368 cubic metres
33.  $23.52